Applied Nanotechnology

The Conversion of Research Results to Products

Second Edition

Applied Nanotechnology

The Conversion of Research Results to Products

Second Edition

Jeremy J. Ramsden

University of Buckingham, Buckingham, UK

AMSTERDAM • BOSTON • HEIDELBERG • LONDON
NEW YORK • OXFORD • PARIS • SAN DIEGO
SAN FRANCISCO • SINGAPORE • SYDNEY • TOKYO

William Andrew is an imprint of Elsevier

William Andrew is an imprint of Elsevier
225 Wyman Street, Waltham, MA 02451, USA
The Boulevard, Langford Lane, Kidlington, Oxford, OX5 1GB, UK
Radarweg 29, PO Box 211, 1000 AE Amsterdam, The Netherlands

Second edition 2014

Library of Congress Cataloging-in-Publication Data
A catalog record for this book is available from the Library of Congress

British Library Cataloguing-in-Publication Data
A catalogue record for this book is available from the British Library

ISBN: 978-1-4557-3189-3

For information on all Academic Press publications
visit our website at http://store.elsevier.com

Printed and bound by CPI Group (UK) Ltd, Croydon, CR0 4YY

Transferred to digital print 2013

No ládd, e nép, mely közt már senki nem hisz,
Ami csodás, hogyan kapkodja mégis.

IMRE MADÁCH

Contents

Preface to the Second Edition xiii
Preface to the First Edition xv

PART I TECHNOLOGY BASICS

CHAPTER 1 What is Nanotechnology? 3
1.1 Nanotechnology as Process 4
1.2 Nanotechnology as Materials 7
1.3 Nanotechnology as Materials, Devices, and Systems 8
1.4 Direct, Indirect, and Conceptual Nanotechnology 9
1.5 Nanobiotechnology and Bionanotechnology 9
1.6 Nanotechnology—Toward a Definition 10
1.7 The Nanoscale 10
1.8 Nanoscience 11
References and Notes 11
Further Reading 12

CHAPTER 2 Science, Technology, and Wealth 13
2.1 Nanotechnology is Different 17
2.2 The Evolution of Technology 18
2.3 The Nature of Wealth and Value 20
2.4 The Social Value of Science 21
References and Notes 23
Further Reading 24

CHAPTER 3 Innovation 25
3.1 The Time Course of Innovation 27
3.2 Creative Destruction 29
3.3 What Drives Development? 32
3.4 Can Innovation be Managed? 33
3.5 The Effect of Maturity 34
3.6 Interaction with Society 34
References and Notes 35
Further Reading 38

CHAPTER 4 Why Nanotechnology? 39
4.1 Miniaturization of Manufacturing Systems 41
4.2 Fabrication 42

4.3 Performance ... 43
4.4 Agile Manufacturing ... 44
References and Notes ... 45
Further Reading ... 46

PART II NANOTECHNOLOGY PRODUCTS

CHAPTER 5 The Nanotechnology Business ... 49
5.1 Nanotechnology Statistics ... 49
5.2 The Total Market ... 50
5.3 The Current Situation ... 52
5.4 Types of Nanotechnology Products ... 54
5.5 Consumer Products ... 55
5.6 The Safety of Nanoproducts ... 57
References and Notes ... 59
Further Reading ... 60

CHAPTER 6 Miscellaneous Applications ... 61
6.1 Noncarbon Materials ... 62
6.2 Carbon-Based Materials ... 63
6.3 Ultraprecision Engineering ... 65
6.4 Aerospace and Automotive Industries ... 66
6.5 Architecture and Construction ... 66
6.6 Catalysis ... 67
6.7 Environment ... 67
6.8 Food ... 69
6.9 Lubricants ... 74
6.10 Metrology—Instrumentation ... 75
6.11 Minerals and Metal Extraction ... 75
6.12 Paper ... 76
6.13 Security ... 77
6.14 Textiles ... 77
References and Notes ... 79
Further Reading ... 81

CHAPTER 7 Energy ... 83
7.1 Energy Harvesting ... 84
7.2 Production and Storage ... 84
7.3 Energy Efficiency ... 91
7.4 Localized Manufacture ... 94
References and Notes ... 94

CHAPTER 8 Information Technologies **97**
8.1 Silicon Microelectronics 98
8.2 Heat Management 98
8.3 Data Storage Technologies 99
8.4 Display Technologies 100
8.5 Molecule or Particle Sensing Technologies 100
References and Notes 100

CHAPTER 9 Health **103**
9.1 Current Activity 104
9.2 Longer-Term Trends 107
9.3 Implanted Devices 107
9.4 Paramedicine 109
References and Notes 109
Further Reading 110

PART III ORGANIZING NANOTECHNOLOGY BUSINESS

CHAPTER 10 The Business Environment **113**
10.1 The Universality of Nanotechnology 113
10.2 The Radical Nature of Nanotechnology 116
10.3 Intellectual Needs 117
10.4 Company–University Collaboration 119
10.5 Clusters 120
10.6 Assessing Demand for Nanotechnology 120
10.7 Technical and Commercial Readiness (Availability) Levels 123
10.8 Predicting Development Timescales 125
10.9 Nanometrology 127
10.10 Standardization of Nanotechnology 129
10.11 Patents 130
References and Notes 132
Further Reading 134

CHAPTER 11 The Fiscal Environment of Nanotechnology **137**
11.1 Sources of Funds 137
11.2 Government Funding 141
11.3 Endogenous Funding 145
11.4 Geographical Differences between Nanotechnology Funding 148
References and Notes 150
Further Reading 151

CHAPTER 12 Regulation 153
References and Notes 156
Further Reading 157

CHAPTER 13 Some Successful and Unsuccessful Nanotechnology Companies 159
13.1 NanoMagnetics 161
13.2 MesoPhotonics 162
13.3 Enact Pharma 163
13.4 Oxonica 163
13.5 NanoCo 164
13.6 Hyperion 164
13.7 CDT 165
13.8 Q-Flo 166
13.9 Owlstone 166
13.10 Generic Business Models 167
References and Notes 168

CHAPTER 14 The Geography of Nanotechnology 169
14.1 Locating Research Partners 171
14.2 Locating Supply Partners 173
14.3 Categories of Countries 173
References and Notes 175
Further Reading 176

CHAPTER 15 Design of Nanotechnology Products 177
15.1 The Challenge of Vastification 177
15.2 Enhancing Traditional Design Routes 178
15.3 Crowdsourcing [7] 179
15.4 Materials Selection 180
References and Notes 180
Further Reading 181

PART IV WIDER AND LONGER-TERM ISSUES

CHAPTER 16 The Future of Nanotechnology 185
16.1 Productive Nanosystems 186
16.2 Self-Assembly 190
16.3 Molecular Electronics 191

16.4 Quantum Computing ... 192
References and Notes ... 192
Further Reading ... 193

CHAPTER 17 Society's Grand Challenges ... 195
17.1 Material Crises ... 195
17.2 Social Crises ... 198
17.3 Is Science Itself in Crisis? ... 198
17.4 Nanotechnology-Specific Challenges ... 199
17.5 Globalization ... 199
17.6 An Integrated Approach ... 200
References and Notes ... 201

CHAPTER 18 Ethics and Nanotechnology ... 203
18.1 Risk, Hazard, and Uncertainty ... 204
18.2 A Rational Basis for Safety Measures ... 205
18.3 Should We Proceed? ... 206
18.4 What about Nanoethics? ... 207
References and Notes ... 208
Further Reading ... 209

Epilog ... 211

Index ... 213

Preface to the Second Edition

During the four years that have elapsed since the first edition of this book was completed, many changes have taken place in the nano landscape. There have been no exceptionally outstanding scientific or technical developments during this interval—although that remarkable nanomaterial graphene was propelled into prominence by the 2010 Nobel Prize for physics—mostly it has been a time of continuous progress on a broad front. On the other hand there has been a dramatic change in the social and economic landscape. The financial crisis had just begun in September 2008 (if we take the collapse of Lehman Brothers in the USA as the marker) and the recession in the United Kingdom began in 2009. The new coalition government, which came to power in 2010, embarked on a rescue program of stringent austerity, which directly cut the availability of public funds for supporting emerging technologies and seems to have adversely affected the readiness of companies to invest in them.

A lackluster services sector, which until recently contributed about three quarters of Britain's gross domestic product, while manufacturing had shrunk to not much more than 10% (about the same as the contribution of the financial services sector, which has recently been rocked by a series of scandals), has led to a new appreciation of the value of having a solid manufacturing base, and it is now government policy to encourage it. In order to compete with the meteoric rise of China as a manufacturing country, it is recognized that to regenerate "old" economies,[1] manufacturing must, however, be placed on a new footing in order to be competitive with lower labor costs elsewhere. This view was shared by other countries. Nevertheless, they had their own crises. While some would assert that the eurozone crisis was inevitable from the start,[2] concrete evidence of grave instability emerged in Greece in late 2009 and during the course of 2010 spread to Ireland, Portugal, and Spain, with Cyprus becoming the latest eurozone country to sail precariously close to default in 2013.

In parallel with these economic crises, the global grand challenges (how to tackle climate change, pollution, energy and resource shortages, and demographic change) have remained in place. Most of them cry out for technology to come to the rescue, and nanotechnology appears to be the perfect answer: atomically precise manufacture should minimize energy and resource use and waste production and provide

[1]This is, of course, not a very accurate term if one looks back over the past few millennia. Until well into the Industrial Revolution the most important manufacturing countries in the world were India and China, a position that they had maintained for almost 2000 years. It was the policy of the British government to suppress indigenous Indian manufacturers in favor of British ones, a policy that was implemented very effectively: by the late 19th century Britain was the world's greatest manufacturing country. It was not long, though, before it was eclipsed by the United States of America, and for most of the 20th century (the exception being during the aftermath of World War II) the volume of German manufactures has exceeded that of Britain, but for the last 50 years Japan has held second place behind the USA (note that Japan entered a long deflationary period in the 1990s, from which it has yet to truly emerge).

[2]See B. Connolly, *The Rotten Heart of Europe*. London: Faber and Faber (1995).

new devices that can be directly used to collect energy from the sun. In essence, nanotechnology promises to do more with less. To give just one concrete example, the transparent conducting indium tin oxide windows presently used in a multitude of electronic devices and reliant on almost-exhausted supplies of indium (exacerbated by efforts to use less of the metal in each device, which has made its recycling more difficult) can be substituted by a percolating network of carbon nanotubes embedded in a polymer (in common with all carbon-based technologies, its deployment sequesters carbon from the atmosphere as a collateral benefit). Nanotechnology represents the cutting edge of the application of science, where Europe and its diaspora, along with Japan, still has a comparative advantage over the rest of the world. Yet, the nano enterprise is advancing falteringly. While some will simply point to the economic crisis rendering unaffordable the continuation of lavish government funding programs, this book will hopefully show that the opportunities have never been greater provided they are addressed in a sensible manner.

The basic structure of the first edition has been retained, but every chapter has been revised and a substantial amount of new material has been added, which has necessitated the appearance of some new chapters, resulting from the expansion of material that was previously fitted into sections within chapters. One of these new chapters deals with the regulation of nanotechnology, which is currently in a state of considerable flux and should be carefully monitored by all those with a stake in the business, coupled with a vigilant readiness to intervene in order to avoid unwanted and unworkable obligations slipping into the statute books.

Jeremy J. Ramsden
The University of Buckingham
April 2013

Preface to the First Edition

This is as much a book about ideas as about facts. It begins (Chapter 1) by explaining—yet again!—what nanotechnology is. For those who feel that this is needless repetition of a well-worn theme, may I at least enter a plea that as more and more people and organizations (latterly the International Standards Organization) engage themselves with the question, the definition is steadily becoming better refined and less ambiguous, and account needs to be taken of these developments.

The focus of this book is *nanotechnology in commerce*; hence, in the first part dealing with basics, Chapter 2 delves into the fascinating relationship between wealth, technology and science. Whereas for millennia we have been accustomed to technology emerging from wealth, and science emerging from technology, nanotechnology exemplifies a new paradigm in which science is in the van of wealth generation.

The emergence of nanotechnology products from underlying science and technology is an instantiation of the process called innovation. The process is important for any high technology; given that nanotechnology not only exemplifies but really epitomizes high technology, the relation between nanotechnology and innovation is of central importance. Its consideration (Chapter 3) fuses technology, economics and social aspects.

Chapter 4 addresses the question "Why might one wish to introduce nanotechnology?" Nanotechnology products may be discontinuous with respect to existing ones in the sense that they are really new, instantiating things that simply did not exist, or were only dreamt about, before the advent of nanotechnology. They may also be a result of *nanification*, decreasing the size of an existing device, or a component of the device, down to the nanoscale. Not every manufactured artifact can be advantageously nanified; this chapter tackles the crucial aspects of when it is technically, and when it is commercially advantageous.

These first four chapters cover Part I of this book. Part II looks at actual nanotechnology products—in effect, defining nanotechnology ostensively. It is divided into four chapters, the first one giving an overview of the entire market, followed by chapters dealing with, respectively, information technology and healthcare, which are the biggest sectors with strong nanotechnology associations; all other applications, including coatings of various kinds, composite materials, energy, agriculture, and so forth, are collected in another chapter.

Part III deals with more specifically commercial, especially financial aspects, and comprises three chapters. The first two are devoted to business models for nanotechnology enterprises. Particular emphasis is placed on the spin-off company, and the rôle of government in promoting nanotechnology is discussed in some detail. The third chapter deals with special problems of designing nano products.

The final part of the book takes a look toward the future, beginning with Productive Nanosystems; that is, what may happen when molecular manufacturing plays a significant rôle in industrial production. The implications of this future state are so profoundly different from what we have been used to the past few centuries that it is worth discussing, even though its advent must be considered a possibility rather than a certainty. There is also a discussion about the likelihood of bottom–up nanofacture (self-assembly) becoming established as an industrial method. The penultimate chapter asks how nanotechnology can contribute to the grand challenges currently facing humanity. It is perhaps unfortunate that insofar as failure to solve these challenges looks as though it will jeopardize the very survival of humanity, they must be considered as threats rather than opportunities, with the corollary that if nanotechnology cannot contribute to solving these problems, then humanity cannot afford the luxury of diverting resources into it. The final chapter is devoted to ethical issues. Whether or not one accepts the existence of a special branch of ethics that may be called "nanoethics," undoubtedly nanotechnology raises a host of issues affecting the lives of every one of us, both individually and collectively, and which cannot be ignored by even the most dispassionate businessperson.

In summary, this book tries to take as complete an overview as possible, not only of the technology itself, but also of its commercial and social context. This view is commensurate with the all-pervasiveness of nanotechnology, and hopefully brings the reader some way toward answering the three questions: What can I know about nanotechnology? What should I do with nanotechnology (how should I deal with it)? What can I hope for from nanotechnology?

Nanotechnology has been and still is associated with a fair share of hyperbole, which sometimes attracts criticism, especially from sober open-minded scientists. But is this hyperbole any different from the exuberance with which Isambard Brunel presented his new Great Western Railway as the first link in a route from London to New York, or Sir Edward Watkin his new Great Central Railway as a route from Manchester to Paris? Moreover, apart from the technology, the nano viewpoint is also an advance in the way of looking at the world; it is a worthy successor to the previous advances of knowledge that have taken place over the past millennium. And especially now, when humanity is facing exceptional threats, an exceptional viewpoint coupled with an exceptional technology might offer the only practical hope for survival.

I should like to especially record my thanks to the members of my research group at Cranfield University, with whom our weekly discussions about these issues helped to hone my ideas, my colleagues at Cranfield for many stimulating exchanges about nanotechnology, and to Dr Graham Holt for his invaluable help in hunting out commercial data. It is also a pleasure to thank Enza Giaracuni for having prepared the drawings.

Jeremy J. Ramsden
Cranfield University
January 2009

PART

I

Technology Basics

CHAPTER

What is Nanotechnology?

1

CHAPTER OUTLINE HEAD

1.1 Nanotechnology as Process 4
1.2 Nanotechnology as Materials 7
1.3 Nanotechnology as Materials, Devices, and Systems 8
1.4 Direct, Indirect, and Conceptual Nanotechnology 9
1.5 Nanobiotechnology and Bionanotechnology 9
1.6 Nanotechnology—Toward a Definition 10
1.7 The Nanoscale 10
1.8 Nanoscience 11

In the heady days of any new, emerging technology, definitions tend to abound and are first documented in reports and journal publications, then slowly get into books and are finally taken up by dictionaries, which do not prescribe, however, but merely record usage. Ultimately the technology will attract the attention of the International Standards Organization (ISO), which may in due course issue a technical specification (TS) prescribing in an unambiguous manner the terminology of the field, which is clearly an essential prerequisite for the formulation of manufacturing standards, the next step in the process.

In this regard, nanotechnology is no different, except that nanotechnology seems to be arriving rather faster than the technologies with which we might be familiar from the past, such as steam engines, telephones, and digital computers. As a reflection of the rapidity of this arrival, the ISO has already (in 2005) set up a Technical Committee (TC 229) devoted to nanotechnologies. Thus, unprecedentedly in the history of the ISO, we shall have technical specifications in advance of the emergence of a significant industrial sector.

The work of TC 229 is not yet complete, however, hence we shall have to make our own attempt to find a consensus definition. As a start, let us look at the roots of the technology. They are widely attributed to Richard Feynman, who in a now famous lecture at Caltech in 1959 [1] advocated manufacturing things at the smallest possible scale, namely atom-by-atom—hence the prefix "nano," atoms typically being a few

Applied Nanotechnology, Second Edition. http://dx.doi.org/10.1016/B978-1-4557-3189-3.00001-4

tenths of a nanometer (10^{-9} m) in size. He was clearly envisaging a manufacturing technology, but from the lecture we also have glimpses of a novel viewpoint, namely that of looking at things at the atomic scale—not only artifacts fashioned by human ingenuity, but also the minute molecular machines grown inside living cells.

1.1 Nanotechnology as process

We see nanotechnology as looking at things—measuring, describing, characterizing, and quantifying them, and ultimately reaching a deeper assessment of their place in the universe. It is also making things. The manufacturing aspect was evidently very much in the mind of the actual inventor of the term "nanotechnology," Norio Taniguchi from the University of Tokyo, who considered it as the inevitable consequence of steadily (exponentially) improving engineering precision (Figure 1.1) [2]. Clearly, the surface finish of a workpiece achieved by grinding it cannot be less rough than atomic roughness, hence nanotechnology must be the endpoint of ultraprecision engineering.

At the same time, improvements in metrology had reached the point where individual atoms at the surface of a piece of material could be imaged, hence visualized on a screen. The possibility was of course already inherent in electron microscopy, which was invented in the 1930s [3], but numerous incremental technical improvements were needed before atomic resolution became attainable. Another development was the invention of the "Topografiner" by scientists at the US National Standards

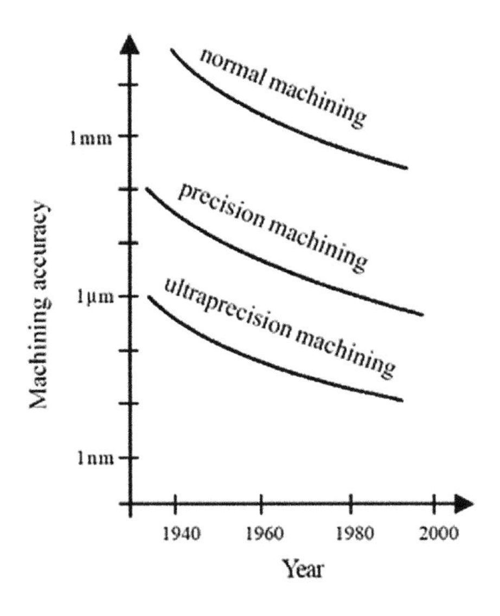

FIGURE 1.1

The evolution of machining accuracy (after Norio Taniguchi).

Institute [4]. This instrument produced a map of topography at the nanoscale by raster scanning a needle over the surface of the sample. A few years later, it was developed into the scanning tunneling microscope (STM), and in turn the atomic force microscope (AFM), which is now seen as the epitome of nanometrology (collectively, these instruments are known as scanning probe microscopes, SPMs). Hence, a little more than 10 years after Feynman's lecture, advances in instrumentation already allowed one to view the hitherto invisible world of the nanoscale in a very graphic fashion. There is a strong appeal in having a small, desktop instrument (such as the AFM) able to probe matter at the atomic scale, which contrasts strongly with the bulk of traditional high-resolution instruments such as the electron microscope, which needs at least a room and perhaps a whole building to house it and its attendant services. Every nanotechnologist should have an SPM in his or her study!

In parallel, people were also thinking about how atom-by-atom assembly might be possible. Erstwhile Caltech colleagues recall Richard Feynman's dismay when William McLellan constructed a minute electric motor by hand-assembling the parts in the manner of a watchmaker, thereby winning the prize Feynman had offered for the first person to create an electrical motor smaller than 1/64th of an inch. Although this is still how nanoscale artifacts are made (but perhaps for not much longer), Feynman's concept was of machines making progressively smaller machines ultimately small enough to manipulate atoms and assemble things *at that scale*. The most indefatigable subsequent champion of that concept was Eric Drexler, who developed the concept of the *assembler*, a tiny machine programmed to build objects atom-by-atom. It was an obvious corollary of the minute size of an assembler that in order to make anything of a size useful for humans, or in useful numbers, there would have to be a great many assemblers working in parallel. Hence, the first task of the assembler would be to build copies of itself, after which they would be set to perform more general assembly tasks.

This program received a significant boost when it was realized that the scanning probe microscope (SPM) could be used not only to determine nanoscale topography, but also as an assembler. IBM researchers iconically demonstrated this application of the SPM by creating the logo of the company in xenon atoms on a nickel surface at 4 K: the tip of the SPM was used to laboriously push 18 individual atoms into location [5]. Given that the assembly of the atoms in two dimensions took almost 24 h of laborious manual manipulation, few people associated the feat with steps on the road to molecular manufacturing. Indeed, since then further progress in realizing an assembler has been painstakingly slow [6]; the next milestone being Oyabu et al.'s demonstration of picking up (abstracting) a silicon atom from silicon surface and placing it at somewhere else on the same surface, and then carrying out the reverse operation [7]. These demonstrations were sufficiently encouraging to stimulate the very necessary parallel work to automate the process of pick-and-place synthesis [8]. Without computer-controlled automation, atom-by-atom assembly could never evolve to become an industrially significant process.

Meanwhile, following on in the spirit of Taniguchi, semiconductor processing—the sequences of material deposition and etching through special masks used to create electronic components [9]—integrated circuits—was steadily reducing the feature sizes that could be achieved well below the threshold of 100 nm that is usually considered to constitute the upper boundary of the nano realm (the lower boundary being about 0.1 nm, the size of atoms). Nevertheless, frustration at being unable to apply "top-down" processing methods to achieve feature sizes in the truly atomic scale (i.e., of the order of 1 nm), or even the tens of nanometers range (although this has now been achieved by the semiconductor industry [10]), stimulated the development of "bottom–up" or self-assembly methods. These were inspired by the ability of randomly ordered structures, or mixtures of components, to form definite structures in biology. Well known examples are proteins (merely upon cooling, a random polypeptide coil of a certain sequence of amino acids will adopt a definite structure), the ribosome, and bacteriophage viruses—a stirred mixture of the constituent components will spontaneously assemble into a functional final structure [11].

At present, a plethora of ingeniously synthesized organic and organometallic compounds capable of spontaneously connecting themselves to form definite structures are available. Very often these follow the hierarchical sequence delineated by A.I. Kitaigorodskii as a guide to the crystallization of organic molecules (the Kitaigorodskii Aufbau Principle, KAP)—the individual molecules first form rods, the rods bundle to form plates, and the plates stack to form a three-dimensional space-filling object. Exemplars in nature include glucose polymerizing to form cellulose molecules, which are bundled to form fibrils, which in turn are stacked and glued with lignin to create wood. Incidentally, this field already had a life of its own, as supramolecular chemistry [12], before nanotechnology focused interest on self-assembly processes.

Molecular manufacturing, the sequences of pick-and-place operations carried out by assemblers, fits in somewhere between these two extremes. Insofar as a minute object is assembled from individual atoms, it might be called "bottom–up." On the other hand, insofar as atoms are selected and positioned by a much larger tool, it

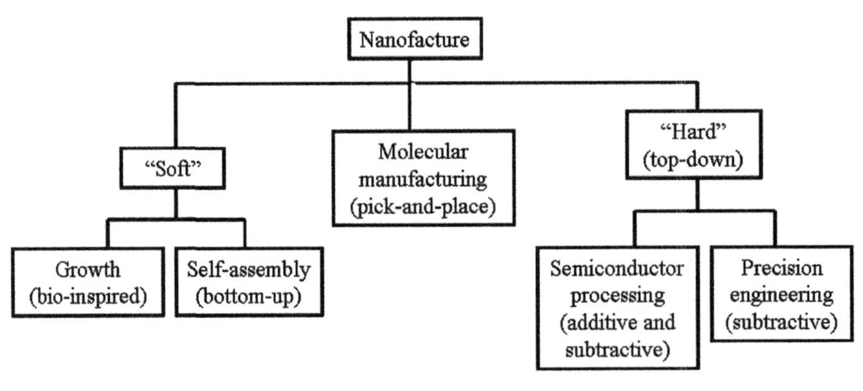

FIGURE 1.2

Different modes of nanomanufacture (nanofacture). "Pick-and-place" assembly is also known as "bottom-to-bottom."

could well be called "top–down." Hence it is sometimes called "bottom-to-bottom." Figure 1.2 summarizes the different approaches to nanofacture (nanomanufacture).

1.2 Nanotechnology as materials

The above illustrates an early preoccupation with nanotechnology as process—a way of making things. Before the semiconductor processing industry reduced the feature sizes of integrated circuit components to less than 100 nm [13], however, there was no real industrial example of nanotechnology at work. On the other hand, while process—top–down and bottom–up, and we include metrology here—is clearly one way of thinking about nanotechnology, there is already a sizable industry involved in making very fine particles, which, because their size is less than 100 nm, might be called nanoparticles. Generalizing, a nano-object is something with at least one spatial (Euclidean) dimension less than 100 nm; from this definition are derived those for nanoplates (one dimension less than a 100 nm), nanofibers (two dimensions less than 100 nm), and nanoparticles (all three dimensions less than 100 nm); nanofibers are in turn subdivided into nanotubes (hollow fibers), nanorods (rigid fibers), and nanowires (conducting fibers).

Although nanoparticles of many different kinds of materials have been made for hundreds of years, one nanomaterial stands out as being rightfully so named, because it was discovered and nanoscopically characterized in the nanotechnology era: graphene and its compactified forms, namely carbon nanotubes (Figure 1.3) and fullerenes (nanoparticles).

A very important application of nanofibers and nanoparticles is in nanocomposites, as described in more detail in Chapter 6.

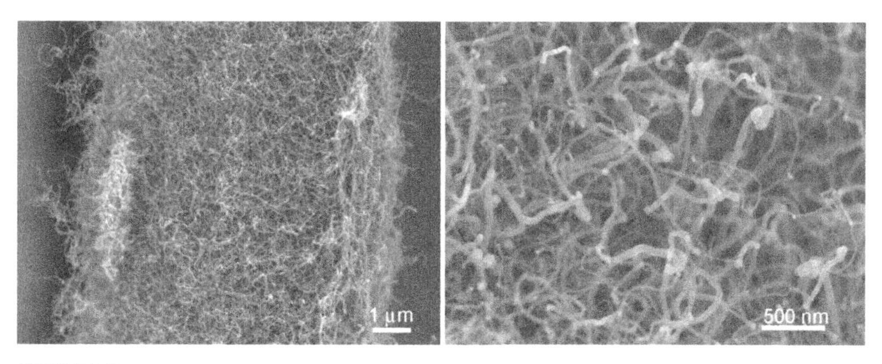

FIGURE 1.3

Scanning electron micrographs of carbon nanotubes grown on the surface of a carbon fiber using thermal chemical vapor deposition. The right-hand image is an enlargement of the surface of the fiber, showing the nanotubes in more detail.

Reprinted from B.O. Boscovic, Carbon nanotubes and nanofibers. Nanotechnol. Perceptions 3 (2007) 141–158, with permission from Collegium Basilea.

1.3 Nanotechnology as materials, devices, and systems

One problem with associating nanotechnology exclusively with materials is that nanoparticles were deliberately made for various esthetic, technological, and medical applications at least 500 years ago, and one would therefore be compelled to say that nanotechnology began then. To avoid that problem, materials are generally grouped with other entities along an axis of increasing complexity, encompassing devices, and systems. A nanodevice, or nanomachine, is defined as a nanoscale automaton (i.e., an information processor), or at least one containing nanosized components. Responsive or "smart" materials could of course also be classified as devices. A device might well be a system (of components) in a formal sense; it is not generally clear what meaning is intended by specifying "nanosystem," as distinct from a device. At any rate, materials may be considered as the most basic category, since devices are obviously made from materials, even though the functional equivalent of a particular device could be realized in different ways, using different materials.

More rigorously than ordering nanotechnology along an axis, these concrete concepts of materials, devices, and systems can be organized into a formal concept system or ontology, as illustrated in Figure 1.4.

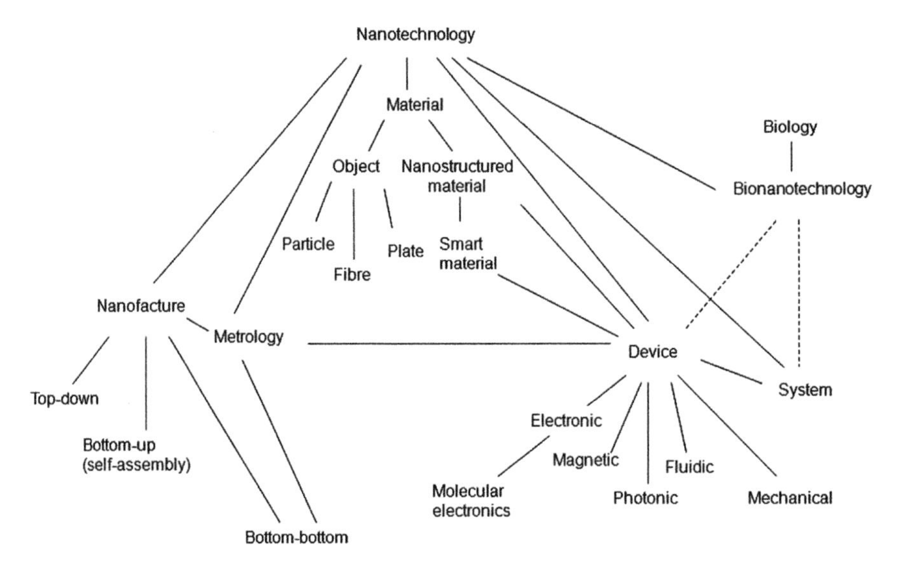

FIGURE 1.4

A concept system (ontology) for nanotechnology. Most of the terms would normally be prefixed by "nano" (e.g., nanometrology, nanodevice). A dashed line signifies that if the superordinate concept contributes, then the prefix must indicate that (e.g., bionanodevice, bionanosystem). Biology may also have some input to nanomanufacture (nanofacture), inspiring, especially, self-assembly processes.

Reproduced from J.J. Ramsden, Towards a concept system for nanotechnology. Nanotechnol. Perceptions 5 (2009) 187–189, with permission of Collegium Basilea.

1.4 Direct, indirect, and conceptual nanotechnology

Another axis for displaying nanotechnology, which might be considered as orthogonal to the materials, devices, and systems axis, considers direct, indirect, and conceptual aspects. Direct nanotechnology refers to nanosized objects used directly in an application—a responsive nanoparticle used to deliver drugs to an internal target in the human body is an example. Indirect nanotechnology refers to a (probably miniature) device that contains a nanodevice, possibly along with other micro or macro components and systems. An example is a cellphone. The internal nanodevice is the "chip"—the integrated electronic information processor circuits with feature sizes less than 100 nm. All the uses to which the cellphone might be put would then also rank as indirect nanotechnology. Given the extent of contemporary society's dependence on information processing, nanotechnology is truly pervasive from this viewpoint alone. It is, of course, the very great processing power, enabled by the vast number of components on a small chip, and the relatively low cost (arising from the same reason), both of which increasingly rely on nanotechnology for their realization, that makes the "micro" processor ubiquitous.

The indirect role of nanotechnology emphasizes its nature as an *enabling technology*. The pervasive presence of computing is enabled by nanotechnology (were processors bigger, they would not be anything like as ubiquitous as they are).

Conceptual nanotechnology refers to considering systems (or, even more generally, "phenomena") from the nano viewpoint—trying to understand the mechanism of a process at the atomic scale. Thus, as an example, molecular medicine, which attempts to explain diseases by the actions of molecules, is classified as conceptual nanotechnology. Conceptual nanotechnology also refers to the creation of novelty, which is exemplified by:

- Objects with no bulk counterpart (e.g., carbon nanotubes).
- Objects acquiring new functionality because of their nano-enabled ultrasmall size or cost (implying the possibility of having vast numbers).
- Enhanced performance, including the achievement of novel combinations of properties by bringing nanoscale components together in intimate juxtaposition.

1.5 Nanobiotechnology and bionanotechnology

These widely used terms are almost self-explanatory. Nanobiotechnology is the application of nanotechnology to biology. For example, the use of semiconductor quantum dots as biomarkers in cell biology research would rank as nanobiotechnology, which encompasses "nanomedicine," defined as the application of nanotechnology to human health.

Bionanotechnology is the application of biology—which could be a living cell, or a biomolecule—to nanotechnology. An example is the use of the protein bacteriorhodopsin as an optically switched optical (nanophotonic) switch.

1.6 Nanotechnology—toward a definition

The current dictionary definition of nanotechnology is "the design, characterization, production and application of materials, devices and systems by controlling shape and size at the nanoscale" [14]. (The nanoscale itself is at present consensually considered to cover the range from about 0.1 to 100 nm—see Section 1.7, but also Ref. [13].) A slightly different nuance is given by the same source as "the deliberate and controlled manipulation, precision placement, measurement, modeling, and production of matter at the nanoscale in order to create materials, devices, and systems with fundamentally new properties and functions." The International Standards Organization (ISO) also gives two meanings: (1) understanding and control of matter and processes at the nanoscale, typically, but not exclusively, below 100 nm in one or more dimensions where the onset of size-dependent phenomena usually enables novel applications; and (2) utilizing the properties of nanoscale materials that differ from the properties of individual atoms, molecules, and bulk matter, to create improved materials, devices, and systems that exploit these new properties. Another formulation encountered in reports is "the design, synthesis, characterization and application of materials, devices, and systems that have a functional organization in at least one dimension on the nanometer scale." The US Foresight Institute gives: "Nanotechnology is a group of emerging technologies in which the structure of matter is controlled at the nanometer scale to produce novel materials and devices that have useful and unique properties." The emphasis on control is particularly important: it is this that distinguishes nanotechnology from chemistry, with which it is often compared; in the latter, motion is essentially uncontrolled and random, within the constraint that it takes place on the potential energy surface of the atoms and molecules under consideration. In order to achieve the desired control, a special, nonrandom *eutactic* environment needs to be available. Reflecting the importance of control, a very succinct definition of nanotechnology is simply "engineering with atomic precision"; sometimes the phrase "atomically precise technologies" (APT) is used to denote nanotechnology; however, we should bear in mind the "fundamentally new (or unique) properties" and "novel" aspects that many nanotechnologists insist upon, wishing to exclude ancient or existing artifacts that happen to be small.

In summary, nanotechnology has three aspects:

1. A universal fabrication procedure.
2. A particular way of conceiving, designing and modeling materials, devices, and systems, including their fabrication (which bears the same relation to classical engineering as "pointillisme" does to classical painting).
3. The creation of novelty.

1.7 The nanoscale

Any definition of nanotechnology must also incorporate, or refer to, a definition of the nanoscale. As yet, there is no formal definition with a rational basis, merely a working proposal. If nanotechnology and nanoscience regard the atom (with size of the order of 1 ångström, i.e., 0.1 nm) as the smallest indivisible entity, this forms a natural lower boundary to the nanoscale. The upper boundary is fixed more arbitrarily. By analogy with microtechnology, now a well-established field dealing with devices up to about 100 μm in size, one could provisionally fix the upper boundary of nanotechnology as 100 nm. However, there is no guarantee that unique properties appear after traversing that boundary from above (see Ref. [13]).

The advent of nanotechnology raises an interesting question about the definition of the prefix "micro." An optical microscope can resolve features of the order of 1 μm in size. It is really a misnomer to also refer to instruments such as the electron microscope and the scanning probe microscope as "microscopes," because they can resolve features at the nanometer scale. It would be more logical to rename these instruments electron nanoscopes and scanning probe nanoscopes—although the word "microscope" is perhaps already too deeply entrenched for a change to be accepted. As a compromise, the term "ultramicroscope" could be used: it is already known within the community of electron microscopists.

1.8 Nanoscience

This term is sometimes defined as "the science underlying nanotechnology"—but is this not biology, chemistry, and physics—or the "molecular sciences?" It is the *technology* of designing and making functional objects at the nanoscale that is new; *science* has long been working at this scale and below. No one is arguing that fundamentally new physics, in the sense of new elementary forces, for example, appears at the nanoscale; rather it is new combinations of phenomena manifesting themselves at that scale that constitute the new technology. The term "nanoscience" therefore appears to be superfluous if it is used in the sense of "the science underlying nanotechnology," although as a synonym of conceptual nanotechnology it might have a valid meaning as the science of mesoscale approximation.

The molecular sciences include the phenomena of life (biology), which do indeed emerge at the nanoscale (although without requiring new elementary laws).

References and notes

[1] Feynman RP. There's plenty of room at the bottom. In: Gilbert HD, editor. Miniaturization. New York: Reinhold; 1961. p. 282–96.
[2] Taniguchi N. On the basic concept of nano-technology. In: Proceedings of the international conference on production engineering Tokyo, Part II (Japan Socisty Precision Engineering); 1974. p. 18–23.
[3] Hawkes PW. From fluorescent patch to picoscopy, one strand in the history of the electron. Nanotechnol Perceptions 2011;7:3–20.
[4] Young R et al. The Topografiner: an instrument for measuring surface microtopography. Rev Sci Instrum 1972;43:999–1011.

[5] Schweizer EK, Eigler DM. Positioning single atoms with a scanning tunneling microscope. Nature (London) 1990;344:524–6.
[6] Apart from intensive activity in numerically simulating the steps of molecular manufacturing—e.g., Temelso B et al. Ab initio thermochemistry of the hydrogenation of hydrocarbon radicals using silicon-, germanium-, tin-, and lead-substituted methane and isobutene. J Phys Chem A 2007;111:8677–88.
[7] Oyabu N et al. Mechanical vertical manipulation of selected single atoms by soft nanoindentation using near contact atomic force microscopy. Phys Rev Lett 2003;90:176102.
[8] Ly DQ et al. The Matter Compiler—towards atomically precise engineering and manufacture. Nanotechnol Perceptions 2011;7:199–217; Woolley RAJ at al. Automated probe microscopy via evolutionary optimization at the atomic scale. Appl Phys Lett 2011;98:253104.
[9] Mamalis AG et al. Micro and nanoprocessing techniques and applications. Nanotechnol Perceptions 2005;1:63–73.
[10] International Technology Roadmap for Semiconductors (ITRS). The current (2009) edition of the ITRS covers developments up to 2025.
[11] Kellenberger E, Assembly in biological systems. In: Polymerization in biological systems, CIBA foundation symposium 7 (new series). Amsterdam: Elsevier; 1972.
[12] Gale PA, Steed JW, editors. Supramolecular chemistry: from molecules to nanomaterials (especially vol. 6: Supramolecular Materials Chemistry). Chichester: Wiley; 2012.
[13] This is a provisional upper limit of the nanoscale. More careful considerations suggest that the nanoscale is, in fact, property dependent. See Ramsden JJ, Freeman J. The nanoscale. Nanotechnol Perceptions 2009;5:3–26.
[14] Abad E et al. NanoDictionary. Basel: Collegium Basilea; 2005.

Further reading

[1] Drexler KE. Engines of creation. New York: Anchor Books/Doubleday; 1986.
[2] Ramsden JJ. What is nanotechnology? Nanotechnol Perceptions 2005;1:3–17.
[3] Ramsden JJ. Nanotechnology: an introduction. Amsterdam: Elsevier; 2011.
[4] Shong CW, Haur SC, Wee ATS. Science at the nanoscale. Singapore: Pan Stanford; 2010.

CHAPTER

Science, Technology, and Wealth

2

CHAPTER OUTLINE HEAD

2.1 Nanotechnology is Different 17
2.2 The Evolution of Technology 18
2.3 The Nature of Wealth and Value 20
2.4 The Social Value of Science 21

Our knowledge about the universe grows year by year. There is a relentless accumulation of facts, many of which are reported in scientific journals, but also at conferences (and which may, or may not, be written down in published conference proceedings) and in reports produced by private companies and government research institutes (including military ones) that may never be published—and some work is now posted directly on the internet, in a preprint archive, or in an online journal, or on a personal or institutional website. The printed realm constitutes the scientific literature [1]. Reliable facts, such as the melting temperature of tungsten, count as unconditional knowledge. Such knowledge does not depend on the particular person who carried out the measurement, or even on human agency (although the actual manner of carrying out the experimental determination depends on both). The criterion of reliability is above all repeatability [2]. These facts are discovered in the same way that Mungo Park discovered the upper reaches of the River Niger.

There is also what is called conditional knowledge: inductive inferences drawn from those facts by creative leaps of human imagination. These are (human) inventions rather than discoveries. Newton's laws (and most laws and theories) fall into this category. They represent, typically, the vast subsuming of pieces of unconditional knowledge into highly compact form. Tables and tables of data giving the positions of the planets in our solar system can be summarized in a few lines of computer code—and the same lines can be used to calculate planetary positions for centuries in the past and to predict them for centuries into the future. Despite the power of this procedure, some people have called it superfluous—the most famous protagonist probably being William of Ockham, whose proverbial razor was designed to cut off all inductive inferences, all theories, not only overly elaborate ones. We must, however,

Applied Nanotechnology, Second Edition. http://dx.doi.org/10.1016/B978-1-4557-3189-3.00002-6

recognize that inductive inference is the heart and soul of science, and John Stuart Mill and others seem to have been close to the truth when they asserted that only inductive, not deductive, knowledge is a real addition to the human store.

There is nothing arcane about the actual description of the theories (although the process by which they are first attained—what we might call the flash of genius—remains a mystery). In the course of an investigation in physics and its relatives, the facts (the primary observations) must first be mapped onto numbers—integer, real, complex, or whatever. This mapping is sometimes called modeling. Newton's model of the solar system, for example, maps a planet with all its multifarious characteristics onto a single number representing a point mass. Then, the publicly accepted rules of mathematics, painstakingly established by generations of mathematicians working out proofs, are used to manipulate those numbers and facilitate the perception of new relations between them.

What motivates this growth of knowledge? Is it innate curiosity, as much a part of human nature as growth in physical stature and mental capabilities? Or is it driven by necessity, to solve problems of daily survival? According to the former explanation, curiosity led to discoveries, which in turn led to practical shortcuts (i.e., technology)—for the production of food in the very early era of human existence and later on for producing the artificial objects that came to be seen as indispensable adjuncts to civilization. Many of these practical shortcuts would involve tools and, later, machines, hence the accumulation of possessions, in other words wealth. As will be discussed in Part III, the whole "machinery" of this process constitutes an indivisible system incorporating also libraries and, nowadays, the internet.

This pattern (Figure 2.1) was later promoted by Francis Bacon in his book *The Advancement of Learning* (1605) with such powerful effect that it thereafter became

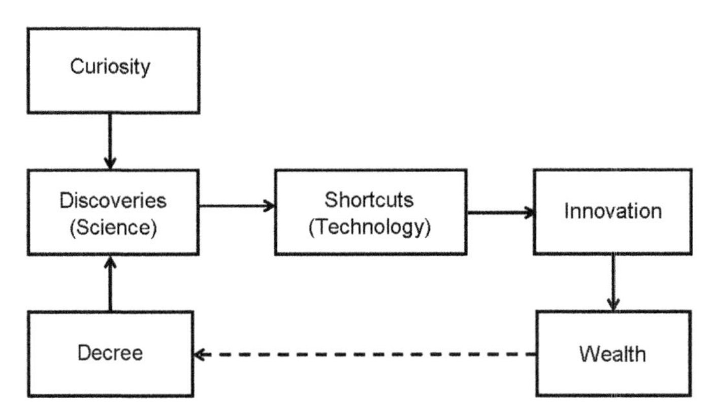

FIGURE 2.1

Sketch of the relationship between science and technology according to the curiosity or decree-driven ("linear") model. Hence, technology can be considered as applied science. The dashed line indicates the process whereby one state, envious of another's wealth, may seek to accelerate the process of discovery.

part of the policy of many governments, remaining so to the present. Bacon was struck by the tremendous political power of Spain in his day. It seemed to heavily preponderate over that of Britain. He ascribed it to technology, which directly resulted from scientific discoveries, which were in turn deliberately fostered (as he believed) by the Spanish government. Nearer our own time, in the Germany of Kaiser Wilhelm, a similar policy was followed (as exemplified most concretely by the foundation of the Kaiser Wilhelm Institutes). In Bacon's mind, as in that of Kaiser Wilhelm, the apotheosis of technology was military supremacy, perceived as the key to political hegemony, the political equivalent of commercial monopoly. Today, the British government, with its apparatus of research councils funding science that must be tied to definite applications with identifiable beneficiaries, is aiming at commercial rather than political advantage for the nation but the basic idea is the same. Similar policies can be found in the USA, Japan, and elsewhere. This model is also known as "linear"; because of the link to government it is also known as the "decree-driven" model.

Bacon's work was published 17 years after the failure of the Spanish Armada, which supposedly triggered his thoughts on the matter. Interestingly, during that interval, although the threat was almost palpable, the feared Counter-Armada never actually materialized. This singular circumstance does not seem to have deflected Bacon from his vision, any more than the failure of Germany's adherence to this so-called "linear model" (science leading directly to technology) to deliver victory in the First World War deflected other governments from subsequently adhering to it. Incidentally, these are just two of the more striking pieces of evidence against that model, which ever since its inception has failed to gather solid empirical support.

The alternative model, which appears to be in much better concord with known facts [3], is that technology, born directly out of the necessity of survival, enables leisure by enhancing productivity, and a small part of this leisure is used for contemplation and scientific activity (Figure 2.2), which might be described as devising ever more sophisticated instruments to discover ever more abstruse facts, modeling those facts, and inferring theories [4]. The motivation for this work seems, typically, to be a mixture of curiosity *per se* and the desire to enhance man's knowledge of his place in the universe. The latter, being akin to philosophy, is sometimes called natural philosophy, a name still used to describe the science faculties in some universities. Those theories might then be used to enhance technology, probably by others than those who invented the theories, enabling further gains in productivity, and hence yet more leisure, and more science. Note that in this model the basic step of creative ingenuity occurs at the level of technology; that is, the practical man confronted with a problem (or simply filled with the desire to minimize effort) hits upon a solution in a flash of inspiration.

A further refinement to this alternative model is the realization that the primary driver for technological innovation is often not linked directly to survival, but is esthetic. Cyril S. Smith has pointed out, adducing a great deal of evidence, that in the development of civilization decorative ceramic figurines preceded cooking utensils, metal jewelery preceded weapons, and so forth [5].

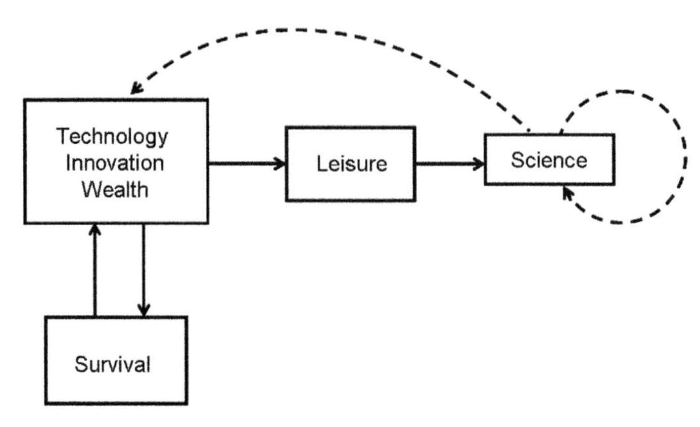

FIGURE 2.2

Sketch of the relationship between science and technology according to the "alternative" model. Technology-enabled increases in productivity allowed Man to spend less than all his waking hours on the sheer necessities of survival. Some part of each day, or month, could be spent in leisure, and while part of this leisure time would be used simply to recuperate from the strains of labor (and should therefore be counted as part of production, perhaps), part was used in contemplation of the world and of events, and sometimes this contemplation would lead to inferential leaps of understanding, adding mightily to knowledge. New knowledge leads to further practical shortcuts, more leisure, and so forth, therefore the development is to some degree autocatalytic (sometimes stated as "knowledge begets knowledge"). The dashed lines indicate positive feedback. According to this view, science can be considered as "applied technology."

Both models adopt the premise that technology leads to wealth. This would be true even without overproduction (i.e., production in excess of immediate requirements), because most technology involves making tools (i.e., capital equipment) that have a relatively permanent existence. Wealth constitutes a survival buffer. Overproduction in a period of plenty allows life to continue in a period of famine. It also allows an activity to be kick-started, rather like the birth of Venus. The Spanish Armada was essentially financed by the vast accumulation of gold and other precious metals from the newly won South American colonies, rather than wealth laboriously accumulated through the working of the linear model, as Bacon imagined.

The corollary is that science cannot exist without wealth. The Industrial Revolution was in full, impressive swing by the time Carnot, Joule, and others made their contributions to the science of thermodynamics. James Watt had no need of thermodynamics to invent his steam engine, although the formal theoretical edifice built up by the scientists later enabled many improvements to be made to the engine [6]. Similarly, electricity was already in wide industrial use by the time the electron was discovered in the Cavendish Laboratory of Cambridge University.

Of course, in society benefits and risks are spread out among the population. Britain accumulated wealth through many diverse industries (Joule's family were brewers, for example). Nowadays, science is almost entirely carried out by a professional corps of scientists, who in the sense of the alternative model (Figure 2.2) spend all their time in leisure; the wealth of society as a whole is sufficient to enable not only this corps to exist, but also to enable it to be appropriately educated—for unlike the creative leaps of imagination leading to practical inventions, the discovery of abstruse facts and the theories inferred from them requires many years of hard study and specialized training, as well as the freedom to think.

2.1 Nanotechnology is different

We can, then, safely assert that all the technological revolutions that have had such profound effects on our civilization (steam engines, electricity, radio, and so forth) began with the technology, and the science (enabled by the luxury of leisure that the technologies enabled) followed later—until the early decades of the 20th century. Radioactivity and atomic (nuclear) fission were purely scientific discoveries, and their technological offshoot, in the form of the atomic pile (the first one was constructed around 1942), was devised by Enrico Fermi, a leading nuclear theoretician, and his colleagues working on the Manhattan project. The rest—nuclear bombs and large-scale electricity-generating plants—is, as they say, history. This "new model," illustrated in Figure 2.3, represents a radical departure from the previous situation. In the light of what we have said above, it begets the question "how is the science paid for?" since it is not linked to any existing wealth-generating activity. The answer

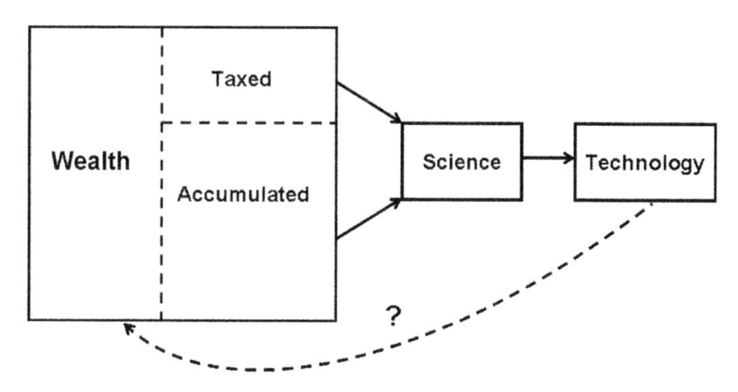

FIGURE 2.3

The "new model" relating wealth, science, and technology, applicable to the nuclear industry and nanotechnology. Note the uncertainty regarding the contribution of these new industries to wealth. There are probably at least as many opponents of the nuclear industry (who would argue that it has led to overall impoverishment; e.g., due to the radioactive waste disposal problem) as supporters. In this respect the potential of nanotechnology is as yet unproven.

appears to be twofold. Firstly, wealth has been steadily accumulating on Earth since the dawn of civilization, and beyond a certain point there is simply enough wealth around to allow one to engage in luxuries such as the scientific exploration of wholly new phenomena without any great concern about affordability. We conjecture that this point was reached at some time early in the 20th century. Secondly, governments acquired (in Britain, largely due to the need to pay for participation in the First World War) an unprecedented ability to gather large quantities of money from the general public through taxation. Since governments are mostly convinced of the validity of the "linear model," science thereafter enjoyed a disproportionately higher share of "leisure wealth" than citizens had shown themselves willing to grant freely in the preceding century. The earlier years of the 20th century also saw the founding of a major state (the USSR) organized along novel lines. As far as science was concerned, it probably represented the apotheosis of the linear model (*qua* "scientific socialism"). Scientific research was seen as a key element in building up technical capability to match that of the Western world, especially the USA. On the whole, the policy was vindicated by a long series of remarkable achievements, especially and significantly in the development of nuclear weapons, which ensured that the USSR acquired superpower status, effectively rivaling the USA even though its industrial base was far smaller [7].

We propose that this "new model" applies to nanotechnology. Several reasons can be adduced in support. One is the invisibility of nanotechnology. Since atoms can only be visualized and manipulated using sophisticated nanoscopes (e.g., scanning probe microscopes) and, hence, do not form part of our everyday experience, they are not likely to form part of any obvious solution to a problem [8]. Another reason is the very large worldwide level of expenditure, and corresponding activity, in the field, even though there is as yet no real nanotechnology industry [9]. In accordance with our insistence (Section 1.6) upon the novelty element needed by any technology wishing to label itself "nano," we do not include the silver halide-based photographic industry and the carbon black (automotive) industry (as will be elaborated in Chapter 5, neither truly fulfils the idea of atomically precise manufacturing).

2.2 The evolution of technology

Human memory, especially "living memory," is strongly biased toward linearity. By far the most common mode of extrapolation into the future is a linear one. Unfortunately for this apparent manifestation of common sense, examination of technology developments over periods longer than the duration of a generation shows that linearity is a quite erroneous perception. Nowadays, there should be little excuse for persistence in holding the linear viewpoint, since most of us have heard about Moore's law, which states that the number of components (transistors, etc.) on an integrated circuit chip doubles approximately every 2 years. This remarkably prescient statement (more an industry prediction than a law) has now held for several decades. But, as Ray Kurzweil has shown, exponential development applies to almost every technology [10]—until, that is, some kind of saturation or fatigue sets in. Of course, any exponential law looks linear provided one examines a short enough interval; that is probably why the linear fallacy persists. Furthermore, at the beginning (of a new technology) an exponential function increases very slowly—and we *are* at the beginning of nanotechnology. Progress—especially in atom-by-atom assembly—is painfully slow at present. On the other hand, progress in information-processing hardware, which nowadays counts as indirect nanotechnology (cf. Section 1.3), is there for all to see. The ENIAC computer (circa 1947) contained of the order of 10^4 electronic components and weighed about 30 tons. A modern high-performance computer capable of 5–10 TFLOPS [11] occupies a similar volume. Formerly, for carrying out large quantities of simple additions, subtractions, multiplications, and divisions, as required in statistics, for example, one might have used the Frieden electromechanical calculator that cost several thousand dollars and weighed several tens of kilograms; the same performance can nowadays be achieved with a pocket electronic calculator costing one dollar and weighing a few tens of grams [12].

The improvements in the performance (speed, energy consumption, reliability, weight, and cost) of computer hardware are remarkable by any standards. If similar improvements could have been achieved with motor-cars, they would nowadays move at a speed of 3000 km/h, use one liter of petrol to travel a 100,000 km, last 10,000 years, weigh 10 mg, and cost about 10 dollars! Comparable improvements in a very wide range of industrial sectors might be achievable with nanotechnology.

Kurzweil (*loc. cit.*) elaborates the exponential growth model applicable to a single technology, placing technology as a whole in the context of the evolution of the universe, in which it occupies one of six epochs:

Epoch 1: Physics and chemistry are dominant; the formation of atomic structures (as the primordial universe, full of photons and plasma, expands and cools).

Epoch 2: Biology emerges; DNA is formed (and, with it, the possibility of replicating and evolving life forms; as far as we know today, this has only occurred on our planet, but there is no principal reason why it could not occur anywhere offering favorable conditions).

Epoch 3: Brains emerge and evolve; information is stored in neural patterns (both in a hard-wired sense and in the soft sense of neural activity; living systems thereby enhance their short-term survivability through adaptability, and hence the possibility of K-selection [13]).

Epoch 4: Technology emerges and evolves; information is stored in artificial hardware and software designs.

Epoch 5: The merger of technology and human intelligence; the methods of biology, including human intelligence, are integrated into the exponentially expanding human technology base. This depends on technology mastering the methods of biology (including human intelligence).

Epoch 6: The awakening of the universe; patterns of matter and energy become saturated with intelligent processes and knowledge; vastly expanded human intelligence, predominantly nonbiological, spreads throughout the universe.

The beginning of Epoch 6 is what Kurzweil calls the singularity, akin to a percolation phase transition.

2.3 The nature of wealth and value

Wealth is defined as accumulated value. A wealthy country is one possessing impressive infrastructure—including hospitals, a postal service, railways, and sophisticated factories for producing goods ministering to the health and comfort of the inhabitants of the country. It also possesses an educated population, having not only universal literacy and numeracy, but also a general interest in intellectual pursuits (as might be exemplified by a lively publishing industry, active theaters and concert halls, *cafés scientifiques* [14], and the like) and a significant section of the population actively engaged in advancing knowledge; libraries, universities, and research institutes also belong to this picture. Thus, wealth has both a tangible, material aspect, and an intangible, spiritual aspect.

This capital—material and spiritual—is, as stated, *accumulated* value. Therefore, we could replace "wealth" in Figures 2.1–2.3 by "value (part of which is refined and accumulated in a store)." We should, therefore, inquire what is value.

Past political economists (such as John Stuart Mill and Adam Smith) have distinguished between value in use and value in exchange (using money). "Value in use" is synonymous with usefulness or utility, perhaps the most fundamental concept in economics, defined by Mill as the capacity to satisfy a desire or serve a purpose. It is actually superfluous to distinguish between value in use and value in exchange, because the latter, equivalent to the monetary value of a good (i.e., its price), is simply a quantitative measure of its value in use. A motivation for making a distinction might have been the "obvious" discrepancies, in some cases, between price and perceived value. But as soon as it is realized that we are only talking about averages, and that the distributions might be very broad, the need for the distinction vanishes. For some individual, a good might seem cheap—to him it is undervalued and a bargain—and for another the converse will be the case. Indeed it might be hard to find someone who values something at exactly the price at which it is offered for sale in the market. A difficulty arises in connection with human life; because there are some ethical grounds for placing infinite value upon it, it might be hard to accommodate in sums. But the insurance industry has solved the problem adequately for the purposes of political economy—it can be equated to anticipated total earnings over a lifetime [15]. A further difficulty arises regarding the possible additional stipulation that for something to have value, there must be some difficulty in its attainment. But here too the difficulty appears to be artificial. Gravity would be more valuable on the Moon than on Earth, where it has, apparently, zero value because it is omnipresent. But perhaps it has zero *net* value: for aviation it is a great nuisance but for motoring it is essential. Air is easily attainable but clean air is a different matter, and even in antiquity whole cities were abandoned because of insufferably bad air. Confusion may arise here because the mode of paying for air is different from that customary for commodities such as copper or sugar. Intrinsically, however, there is nothing terribly arcane about value, which, heuristically at any rate, we can equate with price, and there is not even any need for Pareto's ingenious and more general concept of ophelimity. It should be emphasized that value is always shifting. Certain components of a particular type of aircraft might be very expensive to manufacture, but once that aircraft is no longer in service anywhere in the world, stocks of spare parts become valueless. Mill erred when he tried to determine value relative to some hypothetical fixed standard. The value of almost everything is conditional on the presence of other things, and organized in an exceedingly complicated web of interrelationships.

If utility is considered as the most fundamental concept in economics, the relationship between supply and demand is considered to be the most fundamental law. According to this law, the supply of a good will increase if its price increases, and demand will increase if its price falls, the actual price corresponding to that level of supply exactly matching that of demand—considered to represent a kind of equilibrium. Demand for necessities is stated to be inelastic, because it diminishes rather slightly with increasing price, whereas demand for luxuries is called elastic, because it falls steeply as the price increases. However, this set of relationships has little predictive power. Most suppliers will fix the price of their wares based on a knowledge of the past, and adjustments can be and are constantly being made on the basis of feedback (numbers of units sold) [16]. Because there is a finite supply of many goods (since we live on a finite planet), their supply cannot increase with increasing price indefinitely; on the other hand, the supply of services could in principle be increased indefinitely *pari passu* with demand, which is presumably one of the reasons for the popularity of the "service" rather than the manufacturing economy [17].

There have been numerous attempts to elaborate the simple law of supply and demand. One interesting decomposition of demand is that of Noritaki Kano, into three components: basic, performance, and excitement. For example, the basic needs of the prospective buyer of a motor-car are that it is safe, will self-start reliably, and so forth. Even if the supplier fulfils them to the highest possible degree, the customer will merely be satisfied in a rather neutral fashion, but any deficiency will evoke disappointment. In other words, these attributes are essentially privative in nature. Performance (e.g., fuel consumption per unit distance traveled) typically increases continuously with technological development; customer satisfaction will be neutral if performance is at the level of the industry average; superior performance will evoke positive satisfaction. Finally, if no special effort has been made to address excitement needs (which are not always explicitly expressed, and may indeed only be felt subconsciously), customer satisfaction will be at worst neutral, but highly positive if the needs are addressed. These three components clearly translate directly into components of value.

2.4 The social value of science

Francis Bacon argued in his *Advancement of Learning* (1605) that science discovery should be not just driven by the quest for intellectual enlightenment, but also for the "relief of man's estate." This view is, naturally enough, closely associated with Bacon's "linear" model of wealth creation (Figure 2.1), and forms the basis of the notion (nowadays typically promulgated by state funders of scientific research) that feeding into technological development and wealth creation is an official duty incumbent upon those scientists in receipt of state funds for their work. According to the "alternative model" (Figure 2.2), on the other hand, a scientist voluntarily devotes a part of his or her leisure to research, and there is no especial duty to explicitly promote wealth creation. However, the modern situation of a professional corps of scientists, who are in effect paid by society to devote their whole time to leisure (which they in turn typically wholly devote to research), would appear unarguably to give society the right to demand a specific contribution to the creation of wealth on which, ultimately, the continuation of this arrangement depends. This obligation is most efficiently discharged by insisting on dissemination—an inseparable part of the work of the scientist. Others, with special abilities for wealth creation, can then take up those ideas.

When seeking to analyze the present situation and attempting to present a reasonable recommendation, shifting perspectives during the last few 100 years must be duly taken into account. The Industrial Revolution and the immense wealth it generated managed very well without (or with very little) science feeding into it but, during the last 100 years or so, science has become increasingly associated with obtaining mastery over nature. A survey of the papers published in leading scientific journals shows indeed that a majority is directly concerned with that. However, such work has, in general, been undertaken in a piecemeal fashion. For example, H.E. Hurst's seminal work on the analysis of irregular time series was apparently undertaken at his own initiative while, engaged as Scientific Consultant to the Ministry of Public Works in Egypt, when he was confronted with the need to make useful estimates of the required capacities of the dams being proposed for construction on the Nile. In some cases scientific results were made use of with excellent outcomes; in others with disastrous ones [18]; there are many other examples of both excellence and disasters obtained without any scientific backing. Hence, historical evidence does not allow us to conclude that a scientific research backing guarantees success in a technological endeavor, but rather shows that many other factors, most prominently political ones, intervene. One very positive aspect is that at least this decoupling of science from technology prevented the growth of distortions in the unfettered, disinterested pursuit of objective truth, which almost inevitably becomes a casualty if instead, wealth, is pursued.

But, when it comes to the "new model" (Figure 2.3), we have technology wholly dependent upon science; in other words, without the science there would be no technology and, as already stated, nanotechnology seems to fall into this category. Further implications will be explored in Chapters 3 and 10.

We cannot usefully turn to historical evidence on this point because too little has accumulated. It follows that any extrapolation into the future is likely to be highly speculative. Nevertheless, we cannot rule out the advent of a new era of highly effective science-based handling of affairs that would hopefully yield excellent results. Although the economies and especially the banking sectors of most countries of the world are now rather fragile, to which the response in many circles is rather conservative retrenchment, this is just the wrong kind of response. The whole system of the planet (ecological, social, industrial, financial, and so forth) has been driven so hard to such extremes that mankind can scarcely afford to make more mistakes, in the sense that there is practically no buffering capacity left. Hence, in a very real sense, survival will depend on getting things right. The delicacy of judgment required from the decision-making process is further exacerbated by globalization, thanks to which we now have in effect only one "experiment" under way, and failure means collapse of everything, not just a local perturbation.

References and notes

[1] In passing, it may be noted that this realm is the only one for which consequential source quality appraisal can be carried out. On this point see Wirth W. The end of the scientific manuscript? J Biol Phys Chem 2002;2:67–71.

[2] An important aspect of ensuring the reliability of the scientific literature is the peer review to which reports submitted to reputable scientific journals are subjected. Either the editor himself or a specialist expert to whom the task is entrusted *ad hoc* carefully reads the typescript submitted to the journal and points out internal inconsistencies, inadequate descriptions of procedures, erroneous mathematical derivations, relevant previous work overlooked by the authors, and so forth. The system cannot be said to be perfect—the main weaknesses are: the obvious fact that the reviewer cannot himself or herself actually check the experiments by running them again in his or her laboratory, or verify every step of a lengthy theoretical work, which would take as long as doing the work in the first place; the temptation to undervalue work that contradicts the reviewer's own results; and the pressures imposed by publishers when they are commercial organizations, in which case an additional publishability criterion is whether the paper will sell well, which tends to encourage hyperbole rather than a humbler, more sober, and honest style of investigation. Despite these flaws, it would be difficult to overestimate the importance of the tremendous (and honorary) work carried out by reviewers. This elaborate refining process creates a gulf between the quality of work finally published in a printed journal and the web-based preprint archives, online journals, and other websites. Conference proceedings are in an intermediate position, some papers being reviewed before being accepted for presentation at a conference, but naturally the criteria are different because the primary purpose of a conference is to report work in progress rather than a completed investigation and the discussions of papers represent a major contribution to their value, yet might not even be reported in the proceedings. As regards the internal reports of companies and government research institutes, although they would not necessarily be independently and objectively peer-reviewed in the way that a submission to a journal is, those reports dealing with something of practical value to the organization producing them are unlikely to be a repository of uncertain information, and this provides a kind of internal validation.

[3] Not least the fact that technology has existed for many millennia, whereas science—in its modern sense, as used in all the figures in this chapter—only began in the 12th century CE.

[4] This model is quite similar to one proposed by Adam Smith in his *Wealth of Nations* (1776), Book 5, Chapter 1: industrial money plus old technology enabled new technology to be financed, from which both wealth and academic science sprung.

[5] Smith CS. A search for structure. Cambridge, MA: MIT Press; 1981.

[6] Maxwell's JC paper providing a theoretical (mathematical) foundation for the construction of governors for steam engines, considered to be a landmark (On governors. Proc R Soc 1867–8;16:270–83), appeared almost a century after Watt had actually constructed a working governor.
[7] See Holloway D. Stalin and the bomb. New Haven: Yale University Press; 1994.
[8] It should, however, be borne in mind that these nanoscopes are themselves products of a highly sophisticated *technology*, not science (one may also note that the motivation for developing electron microscopes included a desire to characterize the fine structure of materials used in technological applications).
[9] Apart from an appreciable industry, with a global turnover of around $750 million (electron microscopes and atomic force microscopes), servicing the needs of those developing nanotechnology.
[10] Kurzweil R. The singularity is near. New York: Viking Press; 2005.
[11] 1 TFLOPS is 10^{15} floating-point operations per second.
[12] Assertion of the "same performance" neglects psychology—human factors—the existence of which provides one of the reasons why design is so important (cf. Chapter 15).
[13] See Section 3.1.
[14] They began in Leeds in 1998, modeled on the *café philosophique* started in Paris in 1992, and have become an important forum for debating science issues.
[15] The reader may also recall King James V of Scotland's question "How much am I worth?", which was wittily answered by the miller of Middle Hill as "29 pieces of silver—one less than the value of our Saviour" (Small A. Interesting Roman antiquities recently discovered in Fife. Edinburgh: printed for the author and sold by John Anderson & Co.; 1823).
[16] One of the problems faced by commercial operators is the difficulty of "reading" feedback (let alone responding to it).
[17] Not least since the suppliers of the services mostly themselves require the same services.
[18] The Kongwa (Tanganyika) groundnut scheme of the Overseas Food Corporation serves as an example.

Further reading

[1] Bernal JD. The social function of science. London: Routledge; 1939.
[2] Kealey T. Sex, science and profits. London: Heinemann; 2008.
[3] Mansfield E. Academic research and industrial innovation. Research Policy 1991;20:1–12.
[4] Pethica J, Kealey T, Moriarty P, Ramsden JJ. Is public science a public good? Nanotechnol Perceptions 2008;4:93–112.
[5] Ramsden JJ, Kervalishvili PJ. editors. Complexity and security, especially Chapters 4 and 21. Amsterdam: IOS Press; 2008, for examples of how scientific rationality can be used in policy formulation ("evidence-based sociology").

CHAPTER

Innovation

3

CHAPTER OUTLINE HEAD

3.1 The Time Course of Innovation 27

3.2 Creative Destruction 29

3.3 What Drives Development? 32

3.4 Can Innovation be Managed? 33

3.5 The Effect of Maturity 34

3.6 Interaction with Society 34

Although the dictionary definition of "innovation" is simply "the bringing in of novelties," it has in recent years become a more narrowly defined concept especially liked by government agencies charged with animating economic activity. Indeed, in 2007 the UK government, which has been in the van of this animating process, created a new Department of Innovation, Universities and Skills (merged into the Department for Business, Innovation and Skills created in 2009) [1]. In this usage, innovation has come to mean specifically the process whereby new products are introduced into the commercial sphere: "The technical, designing, manufacturing, management and commercial activities involved in the marketing of a new (or improved) product or the first commercial use of a new (or improved) process or equipment" [2]. It implies not only the commercialization of a major advance in the technological state of the art but also "includes the utilization of even small-scale changes in technological know-how" [3]. Thomas Alva Edison was not only a brilliant inventor but also a masterful innovator (who is reputed to have said "it's 1% inspiration and 99% perspiration"); however, the inventor is very often not the innovator. Suction sweepers are associated not with Spengler, their inventor, but with Hoover; similarly the sewing machine is associated with Isaac Merrit Singer, not with Elias Howe, and still less with Barthélemy Thimonnier or Thomas Saint [4].

The innovator is thus crucial to the overall process of wealth creation. The concept of innovation can be naturally entrained in the "linear model" (Figure 2.1). If we define "high technology" as "technology emerging from science," then nanotechnology is clearly a high technology, according to the "new model" (Figure 2.3) outlined in the

Applied Nanotechnology, Second Edition. http://dx.doi.org/10.1016/B978-1-4557-3189-3.00003-8

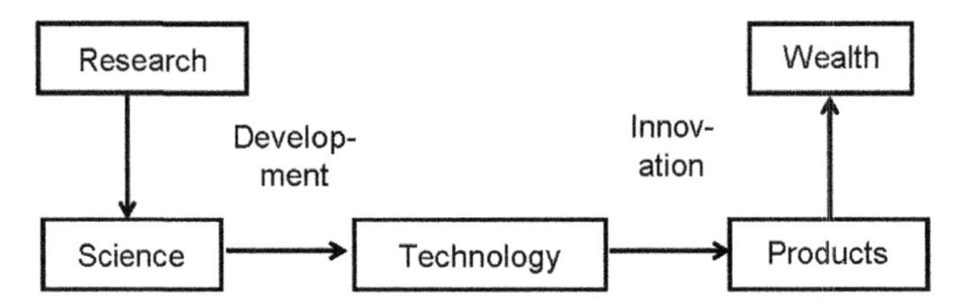

FIGURE 3.1

Detail of the transformation of science to wealth, applicable to both the "linear" and "new" models.

previous chapter, and the process of innovation, in its new constrained usage (of introducing novel products into the commercial sphere), is likely to be highly relevant. Figure 3.1 shows more explicitly how science can be transformed into wealth via innovation.

It is not hard to find reasons for the flurry of official interest in the topic. Governments have noticed that a great deal of research, financed from the public purse, appears to be of very little strategic importance [5]. Even though the funding and execution of scientific research is not, in most countries, prominent in the public mind, nevertheless governments feel that they have to justify public spending on it, despite its negligible contribution to the total government budget [6]. Hence, governments have become obsessed with increasing the economic impact of their science spending [7]. The justification of funding scientific research therefore becomes its capacity to generate wealth through innovation, and the convergence of the "new model" with the "linear model" (although they are not isomorphous) allows the old tradition of Baconian thinking to continue [8].

Innovation, in the sense of the implementation of discovery, or how research results are turned into products, is a theme at the heart of this book (cf. Chapter 10). Governments have become particularly wedded to the path shown in Figure 3.2. Given that the granting of a patent—in other words the right to monopolistically exploit

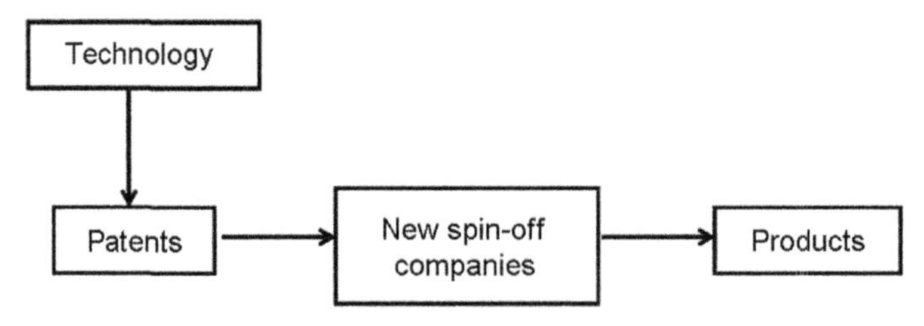

FIGURE 3.2

Detail of the transformation of technology to products, applicable to all the models. New technologies could of course be patented by large established companies as well, but nowadays it is more typical for such companies to buy spin-offs (or start-ups) in order to acquire a new technology.

an invention for a certain number of years—is a clear prerogative of governments, it is perhaps not surprising to find they have a vested interest in promoting patenting, regardless of the presence or absence of any overall economic benefit to the country (cf. Section 10.11); in the USA, the Bayh-Dole Act, which enshrined the right of universities to retain ownership of inventions emerging from federally funded research, was enacted as long ago as 1980.

Economists, especially J.A. Schumpeter, have noticed that established technologies sometimes die out, creating space for new ones. This phenomenon came to be called creative destruction. The man in the street expresses it through proverbs such as "you cannot make an omelette without breaking an egg," and biologists are also familiar with the idea, a good example being the death of about half the neurons at a certain epoch in the development of the brain of the embryonic chicken (and doubtless of other embryonic animals). At the time Schumpeter was putting forward the notion, it was widely believed that epochs of rapid multiplication of new species were preceded by mass destruction of existing ones [9]. Preceding destruction is, however, an unnecessary condition for the occurrence of creative construction. Obviously a literally empty potential habitat has space for colonization (by so-called r-selection—see Section 3.1)—although if it is truly devoid of life initial colonization might be quite difficult. On the other hand, an apparently crowded habitat may be very rich in potential niches for new species capable of imaginatively exploiting them (the so-called K-selection—see Section 3.1). The scientist specializing in biomolecular conformation will be familiar with the fact that for ribonucleic acid (RNA) polymers to adopt their final stable structure, intramolecular bonds formed while the polymer is still being synthesized have to be broken subsequently [10].

Figure 3.1 omits details about the process whereby the new products are transformed into wealth. Evidently, in order for that to happen people must want to buy the products—in other words, there must be a market for them. For incremental technologies, demand for novelty typically comes from buyers of existing products. Directly or indirectly, manufacturers receive feedback from buyers (including the manufacturers' own employees), which can more or less straightforwardly be worked into a steadily improving product. This situation is referred to as "market pull." Disruptive technologies, by definition, are qualitatively different from those in existence at the moment of their emergence. Any user of an existing technology sufficiently farsighted to imagine a qualitatively different solution to his problem is likely himself to be the innovator. Therefore, market pull is inapplicable; one refers to technology push, or the technological imperative. The development of technology is considered to be autonomous, and the emergence of new technologies determines the desire for goods and services [11].

3.1 The time course of innovation

By analogy with biological growth, a good guess for the kinetics would be the sigmoidal logistic equation

$$Q(t) = K/\{1 + \exp[-r(t - m)]\}, \tag{3.1}$$

where Q is the quantity under observation (the degree of innovation, for example), K is the carrying capacity of the system (the value to which Q tends as time $t \to \infty$), r is the growth rate coefficient, and m is the time at which $Q = K/2$. The terms r-selection and K-selection can be explained by reference to this equation: the former operates when an environment is relatively empty and everything is growing as fast as it can (the species with the biggest r will dominate); the latter operates when an ecosystem is crowded (dominance must be achieved by increasing K). This is perhaps more easily seen by noting that Eq. (3.1) is the solution to the differential equation

$$\mathrm{d}Q/\mathrm{d}t = rQ(1 - Q/K). \tag{3.2}$$

The application of this equation to innovation implies, perhaps a little surprisingly, that innovation grows autonomously; that is, it does not need any adjunct (although, as written, it cannot start from zero—we may assume that it begins spontaneously with a lone innovator). Hirooka has gathered some evidence for this time course, the most extensive being for the electronics industry [12]. He promulgates the view that innovation comprises three successive logistic curves: one each for technology, development, and diffusion. ("Development" is used by Hirooka in a sense different from that of Figure 3.1, in which research leads to science (i.e., the accumulation of scientific knowledge) and development of that science leads to technology, out of which innovation creates products such as the personal computer). There seems to be no need to have separate "development" and "diffusion" trajectories: taken together they constitute innovation. In Hirooka's electronics example, the technology trajectory began with the point-contact transistor invented in 1948, and m was reached in about 1960 with the metal oxide-semiconductor transistor and the silicon-based planar integrated circuit. This evidence is not, however, wholly satisfactory, not least because there seems to be a certain arbitrariness in assigning values of Q. Furthermore, why the trajectory should end with submicrometer lithography in 1973 is not clear; the continuation of Moore's law up to the present (and it is anticipated to continue for several more years) implies that we are still in the exponential phase of technological progress. The development trajectory is considered to begin with the UNIX operating system in 1969 and continues with other microprocessors (quantified by the number of components on the processor chip, or the number of memory elements) and operating systems, with m reached in about 1985 with the Apple Macintosh computer; the diffusion trajectory is quantified by the demand for integrated circuits (chips).

Perhaps Hirooka's aim was only to quantify the temporal evolution; at any rate, he does not offer a real explanation of the law that he promulgates, but seems to be more interested in aligning his ideas with those of the empirical business cycles of Kondratiev and others [13]. For insight into what drives the temporal evolution of innovation, one should turn to consideration of the noise inherent in a system (whether socio-economic, biological, mechanical, etc.) [14]. Some of this noise (embodied in random microstates) is amplified up to macroscopic expression [15], providing a potent source of microdiversity. Equation (3.2) should, therefore, be replaced by

$$\mathrm{d}Q/\mathrm{d}t = rQ(1 - Q/K) + \xi(t), \tag{3.3}$$

where ξ is a random noise term (a more complete discussion than we have space for here would examine correlations in the noise). This modification also overcomes the problem that Eq. (3.2) cannot do anything if Q is initially zero.

Amplification of the noise up to macroscopic expression is called by Allen "exploration and experiment." Any system in which mechanisms of exploration and experiment are suppressed is doomed in any environment other than a fixed, unchanging one, although in the short-term exploration and experiment are expensive (they could well be considered as the price of long-term survival).

Recognition of microdiversity as the primary generator of novelty does not in itself provide clues to its kinetics. It may, however, be sufficient to argue from analogy with living systems. By definition, a novelty enters an empty (with respect to the novelty) ecosystem; growth is only limited by the intrinsic growth rate coefficient (the r-limited régime in ecology). Inevitably as the ecosystem gets filled up, crowding constraints prevent exponential growth from continuing.

One may legitimately ask whether the first positive term in Eq. (3.3) should be proportional to Q. Usually, innovation depends on other innovations occurring concurrently. Kurzweil comments that technology can sometimes grow superexponentially. Equation (3.1) should, therefore, only be taken as a provisional starting point. We need to consider that technological growth, $\mathrm{d}Q/\mathrm{d}t$, is proportional to Q^n, and carefully examine whether n is, in fact, greater than unity. For this we also need to work out how to place successive entities in a developing technology on a common scale of the degree of development. How much more developed is the MOS transistor than the *p-n* junction transistor? The complexity of the object might provide a possible quantification, especially via the notion of thermodynamic depth [16]. These matters remain to be investigated further.

During the post-m stage we enter the K-limited régime: survival is now ensured not through outgrowing the competition but through ingenuity in exploiting the highly ramified ecosystem. The filling will itself create some new niches, but eventually the system will become saturated. Even factors such as the fatigue of university professors training the researchers and developers through repeatedly having to expound the same material plays a rôle.

3.2 Creative destruction

The development of the electronics industry is perhaps atypically smooth. Successive technologies usually overlapped the preceding ones, and the industry was generally at pains to ensure compatibility of each technological advance with the preceding one. But innovation is often a much more turbulent affair; one may characterize it using words like discontinuity, disruption, or revolution [17].

An early example of disruptive innovation is the stirrup, which seems to have diffused westwards from China around the eighth century CE, and in Europe was taken up on a large scale by the Franks led by Charles Martel. At a stroke it enabled the horse to be used far more effectively in warfare: with stirrups, a knight could hold a heavy lance, the momentum of the horse would be added to his own, and the lance would be virtually unstoppable by any defenses then current. It also enabled arrows to be fired from a longbow by a mounted rider. This is a good example of technology push—there is no actual evidence that a group of knights sat down and decided this was what they wanted to enable them to fight far more effectively in the saddle. It was a push that was to have far-reaching social consequences. Other armies adopted the innovation, and there was a concomitant increase in defensive technology, including armor and fortified castles. Warfare rapidly became significantly more expensive than hitherto. White has argued that this triggered a revolutionary social change [18]— to support the expense, land was seized by leaders like Martel and distributed to knights in exchange for military service, which they then fulfilled at their own expense. The knights in turn took control over the peasants who lived on the land, cultivating it and raising livestock. In other words, the stirrup led to the introduction of feudalism, a far greater revolution (in the sense that it affected far more people) than that of the technology *per se*.

The classification of innovations as either technology push (typically associated with disruptive innovation: by definition, the market cannot demand something it does not know about) or market pull (for incremental innovations, whereby technology responds to customer feedback) does not seem to cover all cases, however. There is currently no real demand for new operating systems for personal computers, for example, yet Microsoft, for several years the market leader (in terms of volume), is constantly launching new ones. The innovation is incremental, yet customers complain that each successive one is worse than its predecessors (e.g., "Vista" compared with "XP"). Simple economic theory suggests that such products should never be introduced; presumably only the quasi-monopolistic situation of Microsoft allowed it to happen.

An extension to the basic push-pull concept is the idea of "latent demand." It can be identified *post hoc* by unusually rapid takeup of a disruptive innovation. By definition, latent demand is impossible to identify in advance; its existence can only be verified by experiment.

By analogy to supply and demand, push and pull may also (under certain circumstances, the special nature of which needs further inquiry) "equilibrate," as illustrated in Figure 3.3.

As already mentioned near the beginning of this chapter, the term "creative destruction" was introduced by Joseph Schumpeter, but it is in itself incomplete and inadequate for understanding disruptive innovation. It would be more logical to begin with the "noise" (Eq. 3.3), which at the level of the firm is represented by the continuous appearance of new companies—after all, nothing can be destroyed before it exists.

If the commercial *raison d'être* of the company disappears, then the company will presumably also disappear, along with others that depended on it. This process can be modeled very simply: if the firms are all characterized by a single parameter F, which we can call "fitness," and time advances in discrete steps (as is usual in simulations), then at each step the least fit firm is eliminated along with a certain number of its "neighbors" (in the sense of being linked by some kind of commercial dependence)

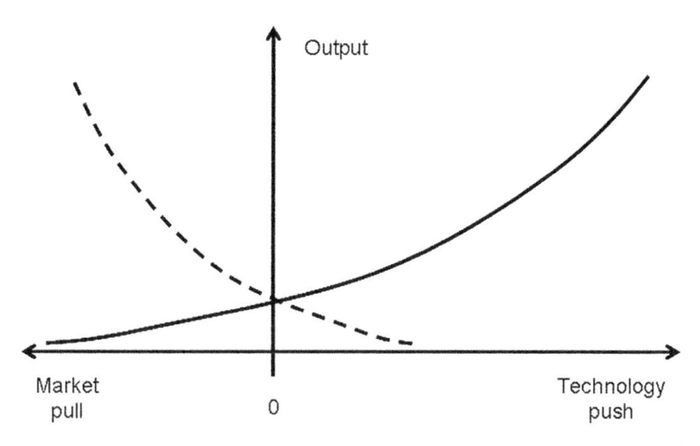

FIGURE 3.3

Proposed quasi-equilibrium between technology push (solid line) and market pull (dashed line). The ideal level of output occurs where they exactly match each other. This diagram neglects consideration of possible temporal mismatch between push and pull.

regardless of their fitnesses, and those eliminated are replaced by new firms with randomly assigned fitnesses [19]. This model was introduced as a description of the evolution of living species, and has some interesting properties, such as a critical (self-organized?) fitness threshold, the height of which depends on the number of neighbors affected by an extinction. Furthermore, if the proper time of the model (the succession of time steps) is mapped onto real (universal or sidereal) time by supposing that the real waiting time for an extinction is proportional to $\exp(F)$ of the eliminated firm, extinctions occur in well-delineated bursts ("avalanches") in real time, whose sizes follow a power law distribution. Paleontologists call this kind of dynamics "punctuated equilibrium" (Figure 3.4) [20].

The Bak–Sneppen model emphasizes the interdependence of species. Companies do not exist in isolation, but form part of an ecosystem. Ramsden and Kiss-Haypál have argued that the "economic ecosystem" (i.e., the economy) optimizes itself such that human desires are supplied with the least effort—a generalization of Zipf's law, whence it follows that the distribution of company sizes obeys [21]:

$$s_k = P(k + \rho)^{-1/\theta}, \tag{3.4}$$

where s_k is the size of the kth company ranked according to size such that k is the rank, $k = 1$ being the largest company, P is a normalizing coefficient, and ρ and θ are the independent parameters of the distribution, called, respectively, the competitive exclusion parameter and the cybernetic temperature. Competitive exclusion means that in any niche, ultimately one player will dominate (this is a simple consequence of general systems theory). The Dixit–Stiglitz model of consumer demand is another application of this generalization [22].

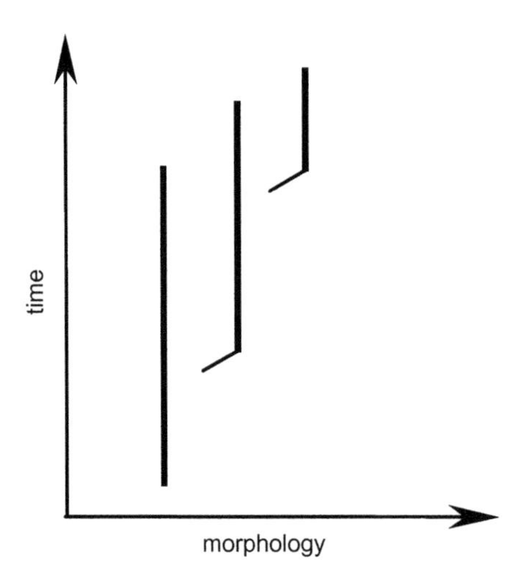

FIGURE 3.4

Sketch of speciation according to the punctuated equilibrium concept. Vertical lines correspond to incremental innovations and diagonal lines to disruptive innovations resulting in a change of technological morphology (quantified e.g. as a Hamming distance).

3.3 What drives development?

Is there any deeper underlying mechanism behind the dynamics represented by Eq. (3.3)? Since invention and innovation are carried out by human beings, one should perhaps look at their underlying motivations. An important principle would appear to be that everyone wants to do a good job—if they get the chance to do so [23]. Even more loftily, the instigators of the grand engineering achievements of the Victorian age "spent their whole energy on devising and superintending the removal of physical obstacles to society's wellfare and development ... the idea of development for its own sake had a place, and in the first instance ... the thought of making man's dwelling place more commodious cast into insignificance anticipations of personal enrichment" [24]. It is only natural for technologists to respond to feedback with incremental innovation. Natural curiosity and the energy to follow it up is sufficient to provide a basis for ξ in Eq. (3.3). But the further growth of innovation, including the actual values of the parameters r and K, depends on other factors, including retarding ones such as inertia and friction. It depends on the fiscal environment [25], because much innovation requires strong concentration of capital. It also depends on "intellectual capital"—knowledge and skills—that must in turn depend in some way on public education. Globalization means that, even more so than ever before, "no country is an island" and, in consequence, it becomes difficult to discern what elements of national policy favor innovation.

3.4 Can innovation be managed?

Given that the average lifetime of a firm is a mere 12 years [26], one might suppose that the directors of even the largest and best-managed companies are constantly afraid of the possibility of sudden extinction. The management literature abounds with exhortations to companies to "remain agile" in order to be able to adapt and survive. What can be offered in the way of specific advice? An excellent general principle is to balance exploration with incremental improvement (in other words, revolution with evolution, respectively) [27]. Merging materials research and manufacturing technologies has also been found to be fruitful in practice, especially within large companies that, formerly, initiated new products with design, then proceeded to materials selection, and finally went to manufacture; now, after merging, these processes take place essentially simultaneously. Sometimes the merging is achieved by the simple tactic of housing designers, engineers, scientists, and production staff within the same building, reversing an earlier trend that, flaunting globalism, often ended up having the units responsible for these different activities in different countries. *That* trend has morphed into crowdsourcing, in which innovative ideas are invited from the entire world internet community (discussed further in Chapter 15).

In recent years there has been much emphasis, especially by governments seeking to justify public expenditure on universities, to academic research as the source of innovation. Nevertheless, careful empirical studies have shown that academic research actually makes a very minor contribution to innovation [28]. We return to this theme in Chapter 11. The rôle of clusters in promoting innovation is discussed in Section 10.5.

In fact, it was already well accepted during the Industrial Revolution that most inventions and innovations came from workmen on the shopfloor. Perhaps religion played a rôle in inspiring them to look at their work, much of which must have been rather repetitive and dull, in a fashion more akin to that of deliberate practice [29], which, as we know from the work of Ericsson et al. [30], is very necessary to achieve expertise; they reckon that about 10,000 h of deliberate practice is needed, which for a 40–50 h working week would be fulfilled after about 5 years.

More formally, a fruitful approach would appear to be to start with an empirical examination of whether ξ in Eq. (3.3) can be correlated with factors such as the percentage of personal income that is saved (and, hence, available for concentrating in large capital enterprises); and the organization of public education in a country.

Such empirical examination can yield surprising results. For example, although the number of patents granted to a company does correlate with its spending on research and development, there is no simple relationship between the spending and corporate performance: in other words, money cannot simply buy effective innovation [31].

New, truly disruptive technologies may require special attention paid to public acceptance. It is widely considered that the failure of companies developing genetically modified (GM) cultivated plant technology to foster open discussion with the public was directly responsible for the subsequent mistrust of foodstuffs derived from GM plants, mistrust that is especially marked in Europe, with huge commercial consequences. Problems associated with the lack of discussion were further exacerbated by the deplorable attitude of many supposedly independently thinking scientists, who often unthinkingly sided with the companies, unwarrantedly (given the paucity of evidence) assuming that ecosystems would not be harmed. Insofar as many scientists working in universities are nowadays dependent on companies for research funding, this attitude came close to venality and did nothing to enhance the reputation of scientists as bastions of disinterested, objective appraisal. Nanotechnologists are now being exhorted to pay heed to those mistakes and ensure that the issues surrounding the introduction of the technology are properly debated openly. There is, in fact, an unprecedented level of public dialog on nanotechnology and, perhaps as a direct consequence, a clear majority of the population seems to be well disposed toward it.

3.5 The effect of maturity

One of the greatest discouragements to the introduction of innovation is a high degree of maturity of some technology. This corresponds to the logistic curve (Eq. 3.1) asymptotically approaching $Q = K$. A good example is the market for prostheses (hip, femur, etc.) Although many surgeons implanting them in patients have innovative ideas about how to improve their design, and new materials are emerging all the time, especially nanocomposites and materials with nanostructured surfaces promoting better assimilation with the host tissue, the existing technology is already at such a high level in general it is extraordinarily difficult to introduce novelty into clinical practice. Unlike electronics, the biomedical field is heavily regulated. In many countries, onerous animal trials must be undertaken before a product is permitted to even be tested on humans. Furthermore, after years of consolidation (cf. Figure 13.2), global supply is dominated by two very large companies, both in the USA.

Another way of looking at this is to consider the market as a complex dynamical system with multiple basins of attraction. As proven by Ashby [32], such a system will inevitably end up stuck in one of its basins and exploration will cease. This is the phenomenon of habituation. The system can only be reset if it receives a severe external (exogenous) shock, which is another way of thinking about "creative destruction" [33].

3.6 Interaction with Society

Manufactured artifacts do not exist in isolation, but are embedded in society and their adoption produces lifestyle changes within it. It has been noted that products with high functionality do not always become diffused and do not necessarily create value (welfare) for society [34]. This situation has been modeled by Ueda and others (Figure 3.5), identifying three classes of value creation [34]:

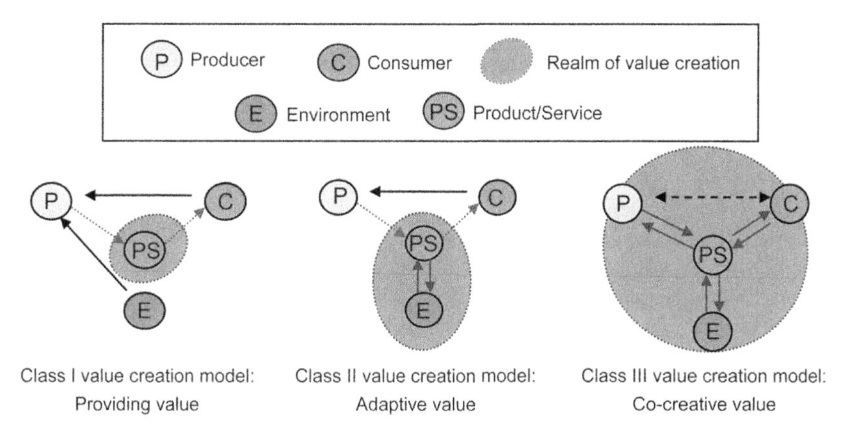

FIGURE 3.5

Value creation model. See text for explanation. From N. Nishino, Co-creative value manufacturing: a methodology for treating interaction and value amongst artifacts and humans in society. Nanotechnol Perceptions 9 (2013) 6–15, with permission from Collegium Basilea.

Value provision (class I)—the value for the producer and the customer can be specified independently and the environment can be determined in advance. The system is closed; it is merely necessary to search for an optimal solution.

Adaptive value (class II)—the value for the producer and the customer can be specified independently but the environment changes. The system is open; an adaptive strategy must be formulated.

Co-created value (class III)—the value for the producer and the customer cannot be specified independently; the two interact and cannot be separated.

It is difficult to find a solution for class III, into which it is expected that most nanoproducts will fall; generally a sophisticated approach such as multi-agent simulation must be used, in which the interactions between humans and artifacts can be explicitly modeled.

References and notes

[1] In December 2011 the Department presented the *Innovation and Research Strategy for Growth* document (Cm 8239) to Parliament.

[2] Freeman C. The economics of industrial innovation. London: Frances Pinter; 1982.

[3] Rothwell R. Successful industrial innovation. R & D Manage 1992;22:221–39.

[4] Thimonnier, indeed, narrowly escaped with his life when tailors smashed his machines in fear of losing their livelihoods.

[5] This is, at root, an indictment of the system of scientific research funded on the basis of the so-called competitive grant proposals. Scientists learn as undergraduates to select

problems of importance to work on and, left to themselves, that is what they will do during their research careers. Unfortunately, the "competitive grant proposal" system does not leave them to themselves. If resources beyond the brain, hand, pencil, and paper are required, funding must be secured by submitting a research proposal to a research council. The research must be described in great detail, and if the proposal is accepted and the research is funded, a contract is drawn up and the scientist is required to exactly follow his or her submitted plans. These may have seemed reasonable at the time the proposal was written, but new knowledge is constantly being discovered and invented, and the scientist is assimilating some of that and having new thoughts of his own, not least while undertaking the research that had been proposed, and the original plan may become irrelevant. The weakness of the system is exacerbated by the slowness of the submission and approval process—roughly 2 years elapses between having the initial idea and starting the research council-funded research—assuming that the proposal is funded (the presently low proportion of proposals that are funded means, of course, that much time—estimates of 40–50% of total working time seemed to be typical nowadays—is wasted writing proposals that never become funded research projects). And, inevitably with this way of organizing things, the report of the completed work delivered to the research council becomes an end in itself. The scientist must be seen to have fulfilled what was contractually required of him, not least since the success of future proposals may depend on the quality of his file maintained by the research council. The actual result is of little consequence (cf. Parkinson CN. In-laws and outlaws. London: John Murray; 1964. p. 134–5.). At meetings convened to review projects, scientists reporting on their work increasingly refer merely to the number of "deliverables" that have been produced—such as the number of papers published, patents applied for or the amount of additional funding secured, with the most cursory attention being given to the actual content of these outputs whereas what one really wants to know is whether new insight and understanding have been generated. When these outputs *are* subjected to closer scrutiny, it often turns out that the basic ideas were published decades ago and then simply forgotten or overlooked. Apart from such duplication, the difficulty of obtaining funding means that there is a strong incentive for the investigator to plod on with his or her research until the end of the contract is reached, even though new knowledge from elsewhere has made the work meaningless.

[6] For example, the funds disbursed by the UK Biotechnology and Biological Sciences Research Council (BBSRC) amount to about one hundredth of government expenditure on health; and the UK's annual contribution to the facilities at CERN, impressive and costly as they are, amounts to a few pounds per head of the population—in other words, a pint or two of beer.

[7] This is explicitly addressed in the "Warry Report"—*Increasing the Economic Impact of Research Councils* (*Advice to the Director General of Science and Innovation, Department of Trade and Industry from the Research Council Economic Impact Group*) (July 2006).

[8] Cf. Gibson I, Silva SRP. Harnessing the full potential of nanotechnology for wealth creation. Nanotechnol Perceptions 2008;4:87–92.

[9] Subsequent, more detailed knowledge of the paleontological record makes this belief untenable; a striking example is the fact that the "Cambrian explosion," perhaps the most remarkable emergence of new species known, was not preceded by a mass extinction event.

[10] Fernández A. Kinetic assembling of the biologically active secondary structure for CAR, the target sequence for the Rev protein of HIV-1. Arch Biochem Biophys 1990;280: 421–4.
[11] See Hodgkinson J et al. Gas sensors 2: the markets and challenges. Nanotechnol Perceptions 2009:5:83–107 for further discussion of this point.
[12] Hirooka M. Complexity in discrete innovation systems. E:CO 2006;8:20–34.
[13] However, a fundamental critique of the cycles is that they fail to take the steady accumulation of knowledge into account. Although the colorful phrase "creative destruction" carries with it the innuendo of *tabula rasa*, of course things are not really like that; although many firms (considered as the basic unit of innovation) are destroyed, the hitherto accumulated knowledge remains virtually intact, because of which history cannot really repeat itself, certainly not to the extent of driving a cycle with unchanging period and amplitude, unless some very special regulatory mechanism is operating (but this is not what is being suggested). In a similar fashion, even though past mass extinctions destroyed up to 90% of all living species, the records of the entire past remained encoded in the DNA of the survivors.
[14] Allen PM, Strathern M, Baldwin JS. Evolutionary drive. E:CO 2006;8:2–19.
[15] Shaw R. Strange attractors, chaotic behaviour, and information flow. Z Naturforsch 1981;36a:80–112.
[16] Lloyd S, Pagels H. Complexity as thermodynamic depth. Ann Phys 1988;188:186–213.
[17] It is as well to remember that whether something is truly revolutionary is a subjective judgment. Clearly the greater the scale and speed of change the more likely it is that something will generally be considered to constitute a revolution, rather than evolution (which need not be uniform).
[18] White L. Medieval technology and social change. New York: Oxford University Press; 1962.
[19] Bak P, Sneppen K. Punctuated equilibrium and criticality in a simple model of evolution. Phys Rev Lett 1993;71:4083–6.
[20] It is especially associated with the names of Ruzhnetsev, Gould, and Vrba.
[21] Ramsden JJ, Kiss-Haypál Gy. Company size distributions in different countries. Physica A 2000;277:220–7.
[22] The relationship (Eq. 3.4) first emerged with a proper derivation in Mandelbrot's work on the analysis of the frequency of word usage in texts (see Mandelbrot BB. Contribution à la théorie mathématique des jeux de communication. Publ Inst Statist Univ Paris 1952;2: 1–124). Later it was applied to systems as diverse as the expression of proteins in bacteria.
[23] Hunter PA, Creating sustained performance improvement. In: Ramsden JJ, Aida S, Kakabadse A, editors. Spiritual motivation: new thinking for business and management, Basingstoke: Palgrave Macmillan; 2007. p. 85–206 [Chapter 15].
[24] Weir A. The historical basis of modern Europe. London: Swan Sonnenschein, Lowrey and Co; 1886:394.
[25] Pethica J. Science: exploration and exploitation. Nanotechnol Perceptions 2008;4:94–7 and cf. [Chapter 11].
[26] Foster R, Kaplan S. Creative destruction. New York: Doubleday; 2001.
[27] Allen PM, Complexity and identity: the evolution of collective self. In: Ramsden JJ, Aida S, Kakabadse A, editors. Spiritual motivation: new thinking for business and management. Basingstoke: Palgrave Macmillan; 2007. p. 50–73 [Chapter 6].

[28] For example, a paper by E. Mansfield (Academic research and industrial innovation, Research Policy 1991;20:1–12) shows that only about 10% of new industrial products and processes emerged from recent academic research; moreover, compared with the 90% resulting from in-house innovations, those originating academically tended to be economically marginal.

[29] Cf. George Herbert's poem "The Elixir" (1633), "Teach me, my God and King/ In all things thee to see … A servant with this clause/ Makes drudgery divine:/ Who sweeps a room, as for thy laws,/ Makes that and th'action fine." The poem was later turned into a hymn by John Wesley (1738), which would doubtless have ensured that the sentiment was well known among the workforces of the early factories. The inspiration for the poem seems to have been an idea of Paul of Tarsus (Letter to the Ephesians 6,7; Letter to the Colossians 3,23).

[30] Ericsson KA, Krampe R Th, Tesch-Romer C. The role of deliberate practice in the acquisition of expert performance. Psychol Rev 1993;100:363–406.

[31] Jaruzelski B, Dehoff K, Bordia R. Smart spenders: the global innovation 1000. McLean, VA: Booz Allen Hamilton; 2006.

[32] Ashby WR. The mechanism of habituation. In: NPL Symposium no 10, mechanization of thought processes. London: HMSO; 1960.

[33] The possibility of intrinsic (endogenous) bursts of destruction should not be neglected, cf. the Bak–Sneppen model (footnote 19).

[34] Nishino N. Co-creative Value Manufacturing: a methodology for treating interaction and value amongst artefacts and humans in society. Nanotechnol Preceptions 2013;9:6–15.

Further reading

[1] Allen PM, Strathern M. Complexity, stability and crises. In: Ramsden JJ, Kervalishvili PJ, editors. Complexity and Security. Amsterdam: IOS Press; 2008:71–92.

[2] Garel G, Mock E. La fabrique de l'innovation. Paris: Dunod; 2012.

[3] Radjou N, Prabhu J, Ahuja S. Jugaad innovation. San Francisco: Jossey-Bass; 2012.

[4] Ramsden JJ. Bioinformatics: an introduction. 2nd ed. London: Springer; 2009—for descriptions of some of the biological concepts useful in discussing innovation, such as *r*- and *K*-selection and the punctuated equilibrium model of the evolution of species.

CHAPTER

Why Nanotechnology?

4

CHAPTER OUTLINE HEAD

4.1 Miniaturization of Manufacturing Systems 41
4.2 Fabrication 42
4.3 Performance 43
4.4 Agile Manufacturing 44

With almost every manufactured product, if the same performance can be achieved by using less material, there will be a cost advantage in doing so. A well-known example is the metal beverage can. Improvements in design—including the formulation of the alloy from which it is made—have led to significantly less material being used for the same function (containing and protecting a beverage). In this example, there are concomitant, secondary advantages of miniaturization (e.g., because the can is also lighter in weight, it costs less to move around). There may be additional issues related to recyclability.

Note that in this case the locus of miniaturization is the thickness of the wall of the can. The basic functional specifications of the can include the volume of beverage that must be contained. This cannot be miniaturized. On the other hand, if the wall could be made of a nanoplate and still fulfil all requirements for strength and impermeability, it would have become a nanoproduct.

In the case of engineering products fulfilling a structural or mechanical purpose, their fundamental scale of size and strength is set by the human being. The standard volume of beverage in a can is presumably based on what a human being likes to drink when quenching his thirst. Perhaps the innovator carried out experiments, much as George Stephenson determined the gauge standard for his railways by measuring the distances between the wheels of a hundred or so farm carts in the neighborhood of Stockton and Darlington and taking the mean, which happened to be $4'8\frac{1}{2}''$ [1].

The length and mechanical strength of a walking stick must be able to support the person using it. Miniaturization of such products therefore generally implies the use of thinner, stronger materials, which might well be nanocomposites, but nevertheless the length of the stick and the dimensions of the hand grip cannot be miniaturized.

Applied Nanotechnology, Second Edition. http://dx.doi.org/10.1016/B978-1-4557-3189-3.00004-X

Another major class of product deals with processing and displaying information. The venerable example is that of the clock, which (in a sense) computes and displays the time of day. Human eyesight places a lower limit on the useful size of the display and other man/machine interfaces for input and output. In the case of mechanical clocks there is a fabrication issue: although a tiny wristwatch uses less material than a standard domestic interior clock, it is more expensive to make, both because the parts must be finished with higher precision and because it is more troublesome to assemble them.

But what is the intrinsic lower limit of the physical embodiment of one bit of information (presence or absence)? Single-electron electronics and Berezin's proposal for isotopic data storage [2] suggest that it is, respectively, one electron or one neutron; in other words one quantum, considered as the irreducible minimum size of matter. But a quantum is absolutely small, in the sense that observing its state will alter it [3]—which seems to suggest that it is useless for the intended purpose. Only in the quantum computer is the possibility that the quantum object can exist in a superposition of states exploited (observation generally forces the elimination of the superposition). Quantum logic therefore implies virtually unlimited parallelism (superposition) in computation, hence an enormous increase in power compared with conventional linear (sequential) logic—provided entanglement with the external environment can be avoided. Although intensive research work is currently being undertaken to develop quantum computers, it has yet to bear fruit in the shape of a working device and, therefore, strictly speaking falls outside the scope of this book, which is focused on actual products.

Conventional logic, in which something is either present or absent and in which the superposition of both presence and absence does not exist, must be embodied in objects larger than individual quanta. The lower size limit of this physical embodiment seems to be a single atom. In principle, therefore, it seems that information storage (memory) could be based on cells capable of containing a single atom, provided what is being observed is not a quantum state, without any loss of functionality.

The most dramatic progress in miniaturization has, therefore, occurred in information storage and processing [4]. In this case, the fabrication technology has undergone a qualitative change since Jack Kilby's first integrated circuit. Making large-scale integrated circuitry in the same way that the first integrated groups of components were made—the mode of the watchmaker—would be prohibitively expensive for a mass-market commodity. Semiconductor processing technology, however, combines miniaturization with parallelization. Not only have the individual components become smaller, but the area processed simultaneously has dramatically increased (measured by the standard diameter of the silicon wafers on which the circuits are built, which has increased from 3 inches up to 12 inches, with 18 inches anticipated).

Within the processor, miniaturization means not only having to use a smaller quantity of costly material, but also shorter distances between components. Since information processing speed is limited by the time taken by the information carriers—electrons—to traverse a component, processing has become significantly faster as a result. Furthermore, since information processing is irreversible, heat is dissipated, and miniaturization also miniaturizes the quantity of heat dissipated per logical operation (although the diminution is less than the increase of density of components executing logical operations, hence managing heat removal is becoming an even greater problem than hitherto). The miniaturization has therefore gone beyond maintaining the same performance using less material, but has actually enhanced performance.

Nevertheless, regardless of the actual sizes of the circuits in which the information processing takes place, the computer/human interface has perforce had to remain roughly the same size. Even so, the nature of the interface has undergone a profound change. Formerly, the processing units were contained in a large room maintained at a fixed temperature and humidity. Job requests were typically handed to an operator, who would load them onto the computer and in due course collect a printout of the results, placing them for pickup by the requester. Miniaturization of the processing units has revolutionized computing in the sense that it has enabled the creation of the *personal* computer. The owner directly feeds instructions into it, and the results are displayed as soon as the computation has finished. The largest parts of the personal computer are typically the keyboard with which instructions are given and the screen on which results are displayed. The miniature processor-enabled personal computer has made computing pervasive and it would be hard to overestimate the social effects of this pervasiveness. It is an excellent example of the qualitative, indirect results of miniaturization.

Another issue is accessibility, which is very size-dependent (for example, in an earlier epoch, children were much in demand as chimney sweeps because they were small enough to clamber up domestic chimneys wielding a broom). The complexity of the circuits required for cellular telephony are such that a hand-held device containing them only became possible with the development of miniaturized, very large-scale integrated circuitry. A similar consideration applies to the development of swallowable medical devices equipped with light sources, a camera, and perhaps even sensors and actuators for drug release (Section 9.1).

In summary, the minute size of integrated circuit components also enables circuits of greater complexity to be devised and realized than would otherwise be possible. In addition, qualitatively different functions may emerge from differently sized devices. There are also secondary advantages of smallness, such as a requirement for smaller test facilities.

4.1 Miniaturization of manufacturing systems

This section takes a brief look at the consequences of realizing the Feynman vision of atomically precise nanoscale machines. They are anticipated to have significantly superior performance to conventional machines. The main features are:

Scaling. Assuming that the speed of linear motion remains constant at all scales, a miniature machine can operate at a much higher frequency compared with macroscopic machines. Hence, throughput expressed as the fraction of the machine's own mass that it can process per unit time (relative throughput) will increase proportionally with diminishing size (assuming that the components handled by the machine are similarly miniaturized). Since mass decreases as the cube of linear dimension, absolute throughput will decrease proportionally to the square of the diminished size. This implies that vast parallelization is required to maintain throughput. The diminished mass also implies negligible inertia and gravitational influence. Stiffness, too, decreases with size, implying that the stiffest possible covalent solids should be used, which is why diamond is a preferred material for the construction of nanoscale machines.

Wear. Covalent bonding between atoms is "digital" (or binary)—either there is a bond or there is not. Hence incremental wear, familiar in large material-processing machines, cannot occur. Instead, there is merely a damage probability, which should be much lower than for a large machine, since it is largely dependent on the stress on a part, which sharply diminishes with decreasing size. Furthermore, manufacturing precision can be retained through multiple generations of manufacturing, just as digitally encoded information can be transmitted and retransmitted many times without loss of fidelity.

Precision. It is well known from macroscale machinery that friction and wear diminish with increasing precision of finishing the parts. Atomic precision is the apotheosis of this trend. The familiar microscopic post-processing machining operations such as grinding and polishing are unnecessary for products assembled atom-by-atom. A further corollary of atomic-scale precision is the great intricacy with which products can readily be made.

Parallelization. In order to make humanly useful quantities of products, atomic assemblers first have to make replicas of themselves. The key parameter is the doubling time. A few dozen doublings suffice for a single nano-assembler to produce 1 kg of copies. This strategy also makes the cost of the machines very low, since the development costs for the first machine can be amortized by all the extra copies.

4.2 Fabrication

Provided performance can be maintained, the smaller a device, the less material is used, leading to cost savings per device. Miniaturization also facilitates massively parallel fabrication procedures—indeed great use has already been made of this possibility in the semiconductor processing industry, as evinced by the continuing validity of Moore's law.

Nanoparticles made from electronically useful materials can be printed at high speed onto a suitable substrate (printed electronics). This innovation enables single-use (disposable) devices, with obvious advantages in applications such as medicine, avoiding the extra work (and expense and resources) of sterilization and the risks of cross-patient infection.

At present, products with features a few nanometers in size (i.e., not quite atomically precise) are not produced by assemblers, nor will they be in the foreseeable future.

Devices such as single-electron transistors have to be made by top–down semiconductor processing technology, with which one-off nanoscale artifacts suitable for testing can be made in the laboratory. Low-cost, high-volume manufacture, however, has the following three requirements [5]:

- A superior prespecified performance achieved with *uniformity, reproducibility and reliability*;
- A high yield to an acceptable tolerance;
- A simulator for both reverse engineering during development and right-first-time design (enabling results to be modified to order).

These requirements are fulfilled by the mainstream micro- (becoming nano-) electronics industry. But is there a lower size limit on what can be got out of devices made using conventional semiconductor processing technology (i.e., epitaxy, lithography, and etching)? Suppose that we are trying to make an array consisting of 3 nm diameter features on a 6 nm pitch [6]. Each layer of each pillar contains about 80 atoms and is formed by adding or removing atoms one at a time by deposition or etching, respectively. These processes can be considered to obey Poisson statistics (i.e., the variance equals the mean). Hence the coefficient of variation (the standard deviation, equal to the square root of the variance, divided by the mean) of the area of the pillars is 12%, whereas the reliable performance of the transistors on a VLSI chip requires a coefficient of variation of less than 2%. *Fundamental statistics intrinsic to the process prevent manufacturability below feature sizes of around* 10 nm. Possibly this statistical barrier could be traversed by fabrication in a eutactic environment (i.e., by assemblers), but as mentioned in the preceding section they are not currently realizable.

Nevertheless, the computational requirements for some applications, such as presenting visual and audio data (i.e., pictures and music), might be able to tolerate circuits less uniform and reliable than those of current VLSI chips [7].

While devices assembled from preformed nano-objects such as carbon nanotubes might not be subject to Poisson statistics (e.g., if they are assembled by a tip-based manipulator—although that is scarcely a high throughput manufacturing technology at present!), another problem then arises—that of the uniformity, reproducibility and reliability of the nano-objects. For example, one of the principal ways of making carbon nanotubes is plasma-enhanced chemical vapor deposition, a process that is evidently subject to Poisson statistics.

4.3 Performance

Performance may be enhanced by reducing the size. If the reason for the size reduction is accessibility or ease of fabrication, the scaling of performance with size must be analyzed to ensure that performance specifications can still be achieved. It is worth noting that the performance of many *micro*systems (microelectromechanical systems, i.e., MEMS) devices actually degrades with further miniaturization [8], and the currently available sizes reflect a compromise between performance and other desired attributes.

If vast quantities of components can be made in parallel very cheaply, devices can be designed to incorporate a certain degree of redundancy, immunizing the system as a whole against malfunction of some of its components [7,9]. Low unit cost and low resources required for operation make redundancy feasible, hence high system reliability. A more advanced approach is to design the circuit such that it can itself detect and switch out faulty components. Note that for malfunctions that depend on the presence of at least one defect (e.g., an unwanted impurity atom) in the material constituting the component, if the defects are spatially distributed at random, the smaller the component, the smaller the fraction that are defective [10].

4.4 Agile manufacturing

Outside the nanoworld, manufacturing is being transformed by digitization. Hitherto, digitization has been closely tied up with information transmission and processing (based on the general-purpose digital computer). Economists have found that economic growth is correlated with the degree of digitization, and a recent refinement of the analysis of the phenomenon suggests that digitization actually causes economic growth [11]. In terms of manufacturing, digitization has meant the computer control of machine tools in factories, in turn permitting the widespread introduction of robots. These factories are engaged in the mass production of identical objects. The concept of additive manufacturing (AM), in which objects are built up by microwelding small lumps of metal (or other materials) to each other, initially developed about 30 years ago for the rapid prototyping of artifacts that would ultimately be manufactured conventionally, has permitted digitization to penetrate more deeply into manufacturing practice: AM (also known as three-dimensional (3D) printing) is now used as a manufacturing technology in its own right. Miniaturized down to atomic dimensions, it would become Feynman–Drexler bottom-to-bottom assembly. A similar trend has occurred in the printing industry: books can now be printed digitally on demand. The traditional economies of scale associated with assembly lines no longer apply: it is possible to make customized objects for practically the same cost as identical ones ("mass customization").

In essence, nanotechnology represents the "digitization" of materials as well as of information, in which objects are fabricated from preformed nanoblocks. By analogy, we can envisage that this further digitization will have a similarly beneficial effect on economic growth. The fundamental reason for these effects of digitization, which does not appear to have been discerned by economists, must be the universalization enabled by digitization. A digital computer can be programmed to carry out any kind of computation; a digital storage medium can store any kind of (digital) information (text, music, or images); and so forth, which in turn enormously increases the adaptability of the system as a whole. Although this might be considered as a revolution, it makes further industrial development easier because new challenges can be met with only a minimal need for brand-new materials; it will suffice to tweak the specifications of a nanoblock and the assembly procedure to create something with quite different properties, for example—the very antithesis of a revolution based on disruptive technology. It also enables everything to be precisely tailored for optimal use. To give just a simple example, at present hip prostheses are provided in about half a dozen standard sizes and the surgeon selects the best match to the patient needing a replacement. The digital concept means that a prosthesis will be rapidly machined using patient data in order to *precisely fit* the body. For the same amount of work, or indeed less work, by the surgeon the result will be far superior, with the tangible economic benefit in terms of the activity of the patient post-operation, as well as other intangible benefits of a superior quality of life.

The nearer nanotechnology approaches the ultimate goal of productive nanosystems (see Section 16.1), the more flexible manufacturing becomes. One aspect that needs careful consideration is the fact that agile (adaptive) production systems are necessarily rooted in algorithms: hence, agile factories must necessarily be computer controlled, due to the large volume of information that has to be processed very rapidly (i.e., in real time) during production. The computer must (at least with present technology) run according to a preloaded program, which is necessarily closed and, hence, cannot but reflect present knowledge. The control center of the factory is therefore intrinsically ill equipped to adapt to the ever-unfolding events that constitute the course of existence, which is largely constituted by the unknowable part of the future. The financial turbulence of 2008, which turned out to have serious industrial consequences, is an all-too-obvious illustration of this truism. Different sectors seem to be intrinsically more or less sensitive to future uncertainty—rational planning demands that this sensitivity be quantified to determine the sectorial appropriateness of agile manufacturing. We need to anticipate that real-world contexts will raise challenges of increasing uncertainty and diversity and so require agility to be achieved by means that are suitably resilient and adaptable to change ("agile agility" or "adaptable adaptability"); that is, agility needs to be explicitly designed for the unknown and unexpected, not merely to cope with well-understood tempos and boundaries. Naturally this will have a cost: presumably the more adaptable an industrial installation, the more expensive; hence it will be appropriate to determine a suitable limit to the necessary adaptability for a particular sector.

References and notes

[1] It is said that Edward Pease, who led the consortium of businessmen promoting the Stockton and Darlington Railway, ordered Stephenson to make the width of the track equal to that of local country carts. This is quite instructive as an example of how *not* to proceed. Railways represented a discontinuity (a disruptive technology) with respect to the technology of farm carts, which was recognized by Stephenson's rival Isambard Brunel, who chose his gauge standard of 7′ by considering the intrinsic possibilities of the new technology. Despite its technical superiority, the reputedly indomitable will of the Stephenson brothers ultimately prevailed, rejecting the Rennie brothers' reasonable

compromise of 5′6″—not only in Britain, but also in much of the rest of the world. It is surprising that the high-speed railways ("shinkansen") in Japan were constructed using the "standard" $4'8\frac{1}{2}''$ gauge, since the existing national railway system had anyway a different gauge of 3′6″; a broader gauge would have allowed even greater speed, stability, and on-board luxury.

[2] Berezin AA. Stable isotopes in nanotechnology. Nanotechnol Perceptions 2009;5:27–36.

[3] Dirac PAM. The principles of quantum mechanics. 4th ed. §§1 and 2. Oxford: Clarendon Press; 1958.

[4] Nevertheless, there is still a long way to go—a memory cell 100 × 100 × 100 nm in size still contains of the order of 10^9 atoms.

[5] Kelly MJ. Nanotechnology and manufacturability. Nanotechnol Perceptions 2011;7: 79–81.

[6] This example is given by Kelly, *loc. cit.*

[7] Lingamneni A et al. Designing energy-efficient arithmetic operators using inexact computing. J Low-Power Electronics 2013;9:141–53.

[8] Hierold C. From micro- to nanosystems: mechanical sensors go nano. J Micromech Microeng 2004;14:S1–11.

[9] See Shannon CE, McCarthy J, editors. Automata studies. Princeton: University Press; 1956.

[10] This is an elementary application of the Poisson distribution. See A. Rényi, Probability Theory. Budapest: Akadémiai Kiadó; 1970. p. 122–5.

[11] Czernich N et al. Broadband infrastructure and economic growth. Economic J 2011;121:505–32.

Further reading

[1] Freitas Jr RA. Economic impact of the personal nanofactory. In: Bostrom N et al, editors. Nanotechnology implications: more essays. Basel: Collegium Basilea; 2006. p. 111–26.

[2] Toth-Fejel T. A few lesser implications of nanofactories. Nanotechnol Perceptions 2009;5:37–59.

PART

NANOTECHNOLOGY PRODUCTS

CHAPTER

The Nanotechnology Business

5

CHAPTER OUTLINE HEAD

5.1 Nanotechnology Statistics 49
5.2 The Total Market 50
5.3 The Current Situation 52
5.4 Types of Nanotechnology Products 54
5.4.1 Products of Substitution 54
5.4.2 Incrementally Improved Products 54
5.4.3 Radically New Products 55
5.5 Consumer Products 55
5.6 The Safety of Nanoproducts 57

This chapter addresses the questions: What nanotechnology is already commercialized? How big is the actual market? How big is the potential market? Coverage of the actual technologies takes place in the three remaining chapters of this part. Here, the main purpose is to put the whole in perspective.

Figure 5.1 summarizes the current situation (cf. Section 1.4). Note that the higher upstream (i.e., closer to the primary source) the nanotechnology, the more indirect the final product. The more indirect, the harder it is to introduce a radical technology, since much more needs to be overturned. Until now, nanotechnology has been most prominent as a substitutional indirect technology (e.g., the introduction of, successively, 65, 45, 32, and 22 nm lithographies in computer chips), and as an incremental quasidirect technology (carriers for active ingredients in cosmetics).

5.1 Nanotechnology statistics

A general caveat is in order here. There is a huge number of statistics about nanotechnology floating around the world. Websites, electronic newsletters, and reports of commercial research are the main secondary sources. Hullman has compiled a summary of some of the secondary sources to create a tertiary report [1], which well highlights the two main (related) problems: the huge variety of numerical estimates for most quantities ("indicators") and the difficulty of defining categories. The main

Applied Nanotechnology, Second Edition. http://dx.doi.org/10.1016/B978-1-4557-3189-3.00005-1

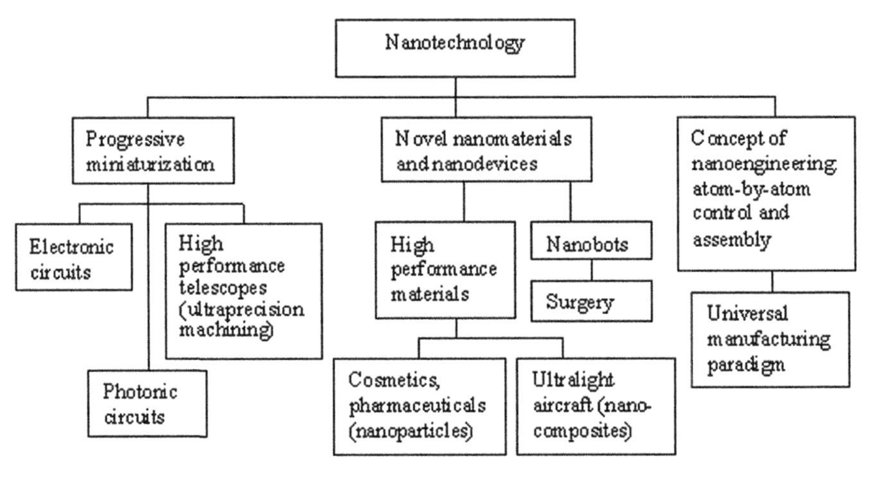

FIGURE 5.1

Indirect, direct, and conceptual branches of nanotechnology (from left to right), with examples (cf. Section 1.4).

reason for the huge discrepancies appears to be the wide variety of definitions of the indicators that are employed. The more easily accessible secondary sources (e.g., electronic newsletters) rarely, if ever, carefully define how they arrive at the quantities given. Reports that are supposed to be based on primary research might be more reliable, but this cannot be established without scrutinizing them in detail, and since they are rather expensive (typically costing several thousand US dollars) few people or organizations acquire a number of different ones and critically compare them. The best solution is probably to undertake the primary research oneself. The nanotechnology industry is still small enough to make this feasible at a cost reasonable compared with that of acquiring the commercial reports, and with a considerable gain in reliability.

Adding to the confusion surrounding the so-called quantitative indicators is the fact that two of the most widely used terms in commercial predictions, "billion" and "trillion", have parallel definitions differing respectively by three and six orders of magnitude from one another. Although the geographical origin of the number and its context usually allow one to reliably decide what is meant, it is regrettable that this ambiguity has been allowed to persist. Usage in the UK is currently the most confusing because, while located in Europe, it uses the same language as the USA. The definitions are summarized in Table 5.1. To avoid confusion, in this book we shall wherever possible write out the numbers explicitly.

5.2 The total market

Figure 1 of the Hullman report (*loc. cit.*) shows the predicted evolution of the world nanotechnology market (presumably defined as sales). Predictions for the year 2010

Table 5.1 Definitions of commonly used words for large numbers.

Quantity	Word Used in:		S.I. Terminology	
	Europe[a]	USA	Prefix	Symbol
10^6	million	million	mega	M
10^9	milliard	billion	giga	G
10^{12}	billion	trillion	tera	T
10^{15}	–	–	peta	P
10^{18}	trillion	–	exa	E

[a] *The same words, with the same meaning, are used in English, French, and German.*

ranged from about $\$10^{11}$ to over $\$1.4 \times 10^{12}$—in other words, roughly the same as the entire present manufacturing turnover in the USA ($\$1.1 \times 10^{12}$ in 2007, one (US) trillion in round numbers; the Taylor Report [2] predicted a global nanotechnology market exceeding $\$2 \times 10^{12}$ around 2012).

A major ambiguity is whether the entire value of a consumer product containing upstream nanotechnology is counted toward the market value; often the nanotechnology only constitutes a small part of the total. Another major ambiguity (see Figure 2 of the Hullman report) is the possibility of double counting. Frequently, the nanotechnology market is divided into different sectors without a clear indication of the criteria for belonging to each division. For example, much of nanobiotechnology is concerned with medical devices, and the devices themselves may contain nanomaterials, yet these are all given separate categories; most aerospace applications involve materials, yet these are two separate categories.

Yet another problem is that it is rarely clear to what extent "old" nanotechnology is included. The world market forecast for nanotechnology given in Figure 5.1 of the Hullman report starts from zero in the year 2001 (but other data given in the same report suggests that in 1999 the world market was already of the order of $\$10^{12}$!). This report is, in fact, unusual insofar as nanotools or nanobiotechnology are given as the dominant sectors (e.g., Figure 3 of the Hullman report) whereas elsewhere it is generally accepted that the overwhelming part of the nanomarket is at present constituted from nanomaterials, with nanodevices and nanotools occupying an almost negligible part [3]. A more reasonable estimate is that the background level of nanomaterials valid at least up to about 2005 is a global turnover of $\$5 \times 10^9$ per annum. This is relatively minor—for comparison, the annual sales of Procter and Gamble in 2007 were $\$75 \times 10^9$. The nanomaterials market is dominated by nanoparticles, which includes (i) a very large volume of carbon black (2006 revenue was approximately $\$1.25 \times 10^9$), chiefly used as an additive for the rubber used to make the tires for motor vehicles; (ii) silver halides used in the photographic industry to sensitize the emulsions with which photographic film is coated (a sector that has sharply declined); and (iii) titanium dioxide used as a white pigment in paint—in other words, "old" nanotechnology (which should therefore not be considered as nanotechnology at all—according to the definitions in Chapter 1).

Furthermore, one almost never sees any uncertainties associated with these estimates. They must often be of the same order of magnitude as the estimates. For example, Figure 12 of the Hullman report shows the distribution of company sizes (measured by turnover) in different countries; but in the USA and the UK the overwhelming majority of companies do not reveal their sizes.

Another criticism is that in many cases a *per capita* comparison would be more relevant than an absolute one; for example, Figure 14 of the Hullman report compares the numbers of institutions active in nanotechnology for European countries—at first glance the graph simply seems to follow the populations of the countries. Even when normalized by population, though, the distribution of institute sizes might vary widely from country to country.

Given these deficiencies, we can only repeat what was already stated above—the best recommendation we can give in this book is that if a company wishes to forecast the market in its particular niche, it had better attempt it by itself—the results are likely to be more reliable than those taken from elsewhere, and the assumptions used in compiling the data can be clearly stated.

5.3 The current situation

In 2003, total global demand for nanoscale materials, tools, and devices was estimated at \$5–8 $\times 10^9$ and forecast to grow at an average annual growth rate (AAGR) of around 30%. The global market value for all of nanotechnology is expected to increase to nearly \$27 $\times 10^9$ in 2015 [4]. The largest segment of the market, namely nanomaterials, is expected to reach nearly \$20 $\times 10^9$ in that year; the second-largest segment, nanotools, a value of about \$7 $\times 10^9$; and the smallest segment, nanodevices, is projected to reach about \$230 million in 2015. Note that in 2007 the largest end-user markets for nanotechnology were environmental remediation (56% of the total market), electronics (21%), and energy (14%). Electronics, biomedical, and consumer applications had much higher projected growth rates than other applications over the following 5 years (30%, 56%, and 46%, respectively). In contrast, energy applications were projected to grow at a CAGR [5] of only 12.5% and environmental applications were predicted to decline by an average of 1.5% per annum. The global market for nanotechnology-*enabled* products is forecast to reach \$2.4 $\times 10^{12}$ by 2015. In 2016, the global market for nanomedicine should reach about \$130 $\times 10^9$, of which anticancer products are expected to comprise about 35% and products for diagnosing disorders of the central nervous system about 23%.

Comparing nanotechnology to other key emerging technologies, the global nanotechnology market is roughly comparable in size to the biotechnology sector, but far smaller than the \$3.6 $\times 10^{12}$ (2012 estimate) global informatics market. However, the nanotechnology market is believed to be growing more than twice as fast as either of the other two.

As stressed in the preceding section, numbers of this nature are necessarily somewhat approximate, not least because of the lack of uniformity regarding the criteria for inclusion; that is, the answer to the question, what is nanotechnology? The global personal computer market is expected to yield revenues (from the microprocessors used in the devices) of about $\$40 \times 10^9$ in 2013; as Moore's law continues its march, most of this market could reasonably be included in the nanotechnology sector on the basis of current device feature sizes.

Nanoscale metal oxide nanoparticles are already very widely used. Typical applications include sunscreens (titanium oxide and zinc oxide), abrasion-resistant coatings, barrier coatings (especially coatings resistant to gas diffusion), antimicrobial coatings, and fuel combustion catalysts. Applications of fullerenes, carbon nanotubes, and a graphene, which are "true" nanotechnology products, still constitute essentially niche markets. In 2008 the fastest-growing nanomaterials segments were nanotubes (with an amazing projected AAGR of 170–180% over the following 5 years) and nanocomposites (about 75% AAGR). These predictions are likely to be upset by the growing prominence of graphene, which is displacing carbon nanotubes in many applications, including composites [6].

The nanomaterials segment, which, as already mentioned, includes several long-established markets ("old" nanotechnology) such as carbon black (used as filler for rubber, especially in automotive applications such as tires), catalytic converter materials, and silver nanoparticles used in photographic films and papers (this sector is now greatly diminished), until fairly recently accounted for almost all (i.e., in excess of 95%) of global nanotechnology sales. By 2008, however, the nanomaterials share of the market had shrunk to around 75% of total sales, and further decline is occurring as nanotools and nanodevices establish a major presence in the market.

The projections naturally depend on a great many imponderables, including the general level of economic activity and economic growth. One highly debatable matter is the influence of government spending. Opinions range from unfavorable (i.e., government spending hinders rather than facilitates technical progress), to favorable (i.e., it makes an indispensable contribution to national competitiveness). We return to this theme in Chapter 10.

In terms of tonnage, total global consumption of all types of nanomaterials was estimated at having surpassed 9 million metric tons in 2005 and 10 million tons in 2010 [7]. Nonpolymer organic materials account for the largest share of total nanomaterials consumption, the bulk of which are carbon black fillers, which is of course a relatively simple traditional material [8]. Metal nanomaterials are the second-largest segment, with more than 20% of the market [9]. The share of simple oxide nanomaterials has been estimated at about 16% in 2010.

Currently hundreds of kinds of nanomaterials are in use or under development, both in their pure form and as composites. Examples include carbon in novel forms [10], tungsten, titanium, and cobalt, as well as many technical ceramics such as new forms of aluminum oxide, silicon carbide, and their composites. Many of these are candidates for adding to paper-based products.

The range of applications for nanomaterials is growing rapidly. Whereas until now nanomaterials have tended to be associated with niche consumer segments such as bouncier tennis balls, a new trend of serious large-scale applications is emerging. These applications currently include tires and other rubber products, pigments, synthetic bone, and automotive components. Tomorrow's applications include automotive coatings, medical devices, and filtration media, to name just a few. Perhaps the ultimate niche application is to the electron sources in electron microscopes, to which the properties of carbon nanotubes are superlative.

One of the difficulties in gathering statistics and appraising the numerous collections that are published almost every month is that it is often not clear exactly what is included as nanotechnology and what is not. For example, carbon black is a traditional material that does indeed have some nano attributes, but it does not belong conceptually to nanotechnology, strictly speaking, because it is not novel. Sometimes what is included within nanotechnology is in effect merely a relabeled traditional product [11].

Nevertheless, very often these near-nano products enhance the attributes of the materials to which they are added to close to the theoretical limit, in which case the almost inevitably higher expense associated with substituting them by real nanomaterials would not result in any increased added value, and hence there is no driver to make the substitution.

5.4 Types of nanotechnology products

5.4.1 Products of substitution

These represent the lowest level of innovation. The consumer may not even be aware of any change; the main advantage is to the producer (lower manufacturing costs through a simplified process or design), and possibly to the environment (a smaller burden, due to the use of a smaller quantity of raw materials, hence less weight to transport and less waste to ultimately be discarded). In this case the anticipated market is the same as the present market; if there is an increasing or decreasing trend it may be considered to continue (e.g., exponential, linear, or logarithmic or a combination of all three, i.e., logistic) in the same fashion.

If the innovation reduces production costs, the enhanced profitability may attract other manufacturers (assuming that the innovation is not protected by patent or secrecy), which would tend to depress the price in the long term.

5.4.2 Incrementally improved products

Examples are tennis rackets reinforced with carbon nanotubes, making them stronger for the same weight. Very often this will make the product more expensive, so elasticity of demand is a significant factor. On the other hand, it is doubtful whether the laborious compilations of demand elasticity that have been made in the past are really useful. What degree of improvement ranks as incremental? It might not take very much for the product to be considered as essentially new. Furthermore, how is one to quantify quality? If a laptop computer originally weighing 2 kg can be made to weigh only 1.5 kg with the same information-processing performance, different users will value the improvement in different ways.

5.4.3 Radically new products

These are goods that, in their qualitative nature, did not exist before. Of course, it is perhaps impossible for something to be totally new. Polaroid "instant" film (that could be developed and made visible seconds after taking a snapshot) was certainly a radical concept, but on the other hand it was still based on a silver halide emulsion and the mode of actually snapping the shot was the same, essentially, as with a Kodak box camera.

The future is in this case very difficult to predict, and an *ad hoc* model (Section 10.6.1) is probably needed if any serious attempt at planning is to be made.

5.5 Consumer products

The Woodrow Wilson Center created a list of consumer products containing nanotechnology that continued to be curated until March 2011, when it listed 1317 products [12]. Some of these are given in the next three tables. This data provides a useful snapshot of the commercial market.

Regarding Table 5.2, it is not always clear what the exact criteria for inclusion are, especially for products that could fit into multiple categories. For example, would a household appliance be included under "Appliances" or under "Home and garden"? Most appliances include some electronics, one imagines. And does an automobile (which may contain some on-board information processors with nanoscale features in their chips) count as a single product? Spray paint containing nanoparticles for use by owners to repair minor scratches presumably ranks as an automotive product, but

Table 5.2 Numbers of consumer products in different categories (status in January 2009).[a]

Category	Number	%
Health and fitness	502	
Home and garden	91	
Food	80	
Electronics	56	
Automotive	43	
Appliances	31	
Other	70	
Total	873	100

[a] *Source: see footnote [12].*

Table 5.3 Numbers of consumer products in the "health and fitness" category (status in January 2009).[a]

Subcategory	Number	%
Personal care	153	
Cosmetics	126	
Clothing	115	
Sporting goods	82	
Filtration	40	
Sunscreen	33	
Total		100

[a] *Source: see footnote [12].*

does each available color count as a separate product? Furthermore, the compilers of the data have not themselves verified whether the manufacturers' claims are correct. Moreover, one has no indication of the volumes sold: cellphones probably outrank all the other members of its category, for example.

It is perhaps surprising that there are already so many food products at least containing, if not based on, nanotechnology—these, incidentally, might well have been included in the "health and fitness" category. The list is anyway dominated by health and fitness products, which are further broken down in Table 5.3. Presumably medical products not available uncontrolled to consumers (e.g., prescription drug delivery nanomaterials) are not included in the list at all—or else none are currently available.

Finally, it is interesting to look at which elements dominate nanotechnology applications (Table 5.4). Presumably these are mostly in the form of nanoparticles. Carbon presumably means fullerenes or CNT. Silicon, titanium, and zinc are presumably nanoparticles of their oxides. Since the database is of consumer products, presumably silicon-based integrated circuits are not included.

Table 5.4 Numbers of consumer products categorized according to the elements declared as constituting the nanocomponent (status in January 2009).[a]

Element	Number of Products	%
Silver	235	
Carbon	71	
Titanium	38	
Silicon	31	
Zinc	29	
Gold	16	
Total		100

[a] *Source: see footnote [12].*

The consumer market is of course extremely fickle. The epithet "nano" is sometimes used as a marketing ploy, even if the product contains no nanomaterials at all [13].

Furthermore, it is evolving with amazing rapidity. A camera in 1960 contained no electronics, but now contains probably 80% or more, much of which is heading toward the nanoscale. A similar trend has occurred regarding personal calculators, the functional equivalent of which would have been a slide rule or a mechanical device in 1960. The personal computer did not even exist then. A motor-car typically contained about 10% (in value) of electronics in 1960; this figure is now between 30% and 50% and much of it is already, or fast becoming, nano.

A crucial point regarding consumer market volume is the renewal cycle. Whereas in other markets technical considerations dominate—for example, in many European cities the underground railway trains and trams might be of the order of 50 years old and still in good working order—psychosocial factors dominate the decision whether to replace a consumer product. It seems remarkable that those who have a mobile phone (i.e., the majority of the population) typically acquire a new one every 6 months (many are anyway lost or stolen). Other consumer electronics items such as a personal computer, video recorder, or television receiver might be renewed every 1–2 years. Even a motor-car is likely to be changed at least every 5 years, despite the many technological advances that ensure that it is still in perfect working order at that age.

Here, deeper issues are raised. Without the frenetic pace of renewal, the hugely expensive infrastructure (e.g., semiconductor processing plants) supporting present technology could not be sustained, and though rapid "planned obsolescence" seems wasteful, without it innovation might grind to a halt, with possibly deleterious consequences for mankind's general ability to meet future challenges (including those associated with global warming).

5.6 The safety of nanoproducts

One issue that has not so far received much prominence is that of safety, especially regarding nanoparticles in products brought into contact with the skin, if not actually ingested. Compared with the furore over genetically modified food crops, leading to widespread prohibition of their cultivation, at least in Europe, nanoparticle-containing products have generally had a favorable reception, perhaps because of the considerable care taken by the industry to inform members of the public about the technological developments that led to them.

Nevertheless, there is no doubt that nanoparticles have significant biological effects. An extensive literature already exists [14]. A member of the public might wish to take the following widely known facts into account:

1. Workers, especially miners, exposed to fine particles suffer occupational diseases such as silicosis and asbestosis. Tumors typically first appear after many years of exposure, and are generally painful, incurable, and fatal.

2. Widespread use of coal for domestic heating (e.g., in London up to the 1950s and in Germany up to the 1990s) led to severe atmospheric pollution and widespread respiratory complaints.
3. On the other hand, restricted exposure to dusts (speleotherapy, e.g., as practiced in the "Rehabilitation" Scientific-Medical Center of the Ukrainian Health Ministry in Uzhgorod (Ungvár)) is considered to be therapeutic; from 1864 to 1905 (when electric traction was introduced) people suffering from respiratory complaints were encouraged to travel on the Metropolitan and District Railways in London, in the days when their trains were still steam-hauled, and hence the tunnels through which they passed were rich in sulfurous fumes.
4. Cigarette smoking in public places is now subject to draconian restrictions (at least in Europe and the USA).
5. The increase of motor traffic in major cities, coupled with official encouragement for diesel engines, which emit large quantities of nanoparticulate carbon in their exhaust, have made air pollution as bad nowadays as it was in the days when domestic heating using coal was widespread.

This list could be prolonged, but the point is made that no coherent policy can be discerned at present; the situation is full of paradoxes. Items 1 to 3 can perhaps be resolved by recalling Paracelsus' dictum "The poison is in the dose", but in the other cases probably economic and political factors took precedence over scientific and medical ones. The British government now seems to be resolved to bring some order into this chaos, and commissioned a report prescribing how studies to determine the biological hazards of nanoparticles ought to be carried out [15]. The dispassionate observer of the field will find it remarkable that, despite decades of investigations, most reported studies have failed to carry out requisite controls, or are deficient in other regards. A very great difficulty of the field is the extremely long incubation time (decades) of some of the diseases associated with exposure to particles. The effects of long-term chronic exposure might be particularly difficult to establish. At the same time, ever since Prometheus man has been exposed to smoke, an almost inevitable accompaniment to fire, and doubtless the immune system has developed the ability to cope with many kinds of particles [16].

The most appropriate response is to make good the deficiencies of previous work, as recommended by Tran et al. (*loc. cit.*) [17]. The question remains, what are we to do meanwhile, since it might be many years before reasonably definitive answers are available. Most suppliers of nanomaterials would, naturally enough, prefer the *status quo* to continue until there is clear evidence for acting otherwise; "pressure groups" are active in promulgating the opposite extreme, advocating application of the precautionary (or "White Queen") principle (do nothing unless it is demonstrably safe) and an innovation-stifling regulatory régime (cf. Chapter 12). The latter is anyway supported by governments (probably, even doing the research required to establish the safety or otherwise of nanoparticles contravenes existing health and safety legislation) and supergovernmental organizations such as the European Commission.

Hence, the most sensible course that can be taken by the individual consumer is to apply the time-honored principle of *caveat emptor*. But, the consumer will say, we are not experts, how can we judge? But, the expert may respond, we all live in a technologically advanced society, and we all have a corresponding responsibility to acquaint ourselves with the common fund of knowledge about our world in order to ensure a long and healthy life. Naturally we have a right to demand that this knowledge is available in accessible and intelligible form.

The only rational way to proceed is to build up knowledge that can then be applied to calculate the risks and weigh them against the possible benefits. Provided the knowledge is there, this can be done quite objectively and reliably (see Section 18.2), but gaining the knowledge is likely to be a laborious task, especially when it comes to assessing the chronic effects resulting from many years of low-level exposure. There is particular anxiety regarding the addition of small metallic or metal oxide nanoparticles to food. Although a lot about their biological effects is indeed already known [14], the matter is complex enough for the ultimate fates of such particles in human bodies to be still rather poorly understood, and new types of nanoparticles are being made all the time [18]. On the other hand, it is also worth bearing in mind that some kinds of nanoparticles have been around for a long time—volcanoes and forest fires generate vast quantities of dust and smoke, virus particles are generally within the nanorange, comestible biological fluids such as milk contain soft nanoparticles, and so forth, merely considering natural sources. Anthropogenic sources include combustion in many forms, ranging from candles, oil lamps, tallow dips, etc., used for indoor lighting, internal combustion engines—this is a major source of nanoparticle pollution in cities, along with the dust generated from demolishing buildings—cooking operations, and recreational smoking. The occupational hazards from certain industries, especially mining and mineral processing (silicosis, asbestosis), are well recognized, and the physicochemical and immunological aspects of the hazards of the nanoparticles are reasonably well understood [19].

References and notes

[1] Hullman A. The Economic Development of Nanotechnology—An Indicators-Based Analysis. Directorate-General for Research, Nano Science and Technology Unit, Brussels: European Commission; 2006.

[2] New Dimensions for Manufacturing: A UK Strategy for Nanotechnology (report of the Advisory Group on Nanotechnology Applications, chaired by John Taylor). London; 2002.

[3] The Hullman report merely compiles secondary sources, without any criticism or even the highlighting of discrepancies.

[4] This prediction, and the numbers following, were issued by the Economic Research Unit of the INSCX exchange (inscx.com) on 2 August 2012.

[5] Compound annual growth rate (CAGR) is the geometric mean growth rate over a period of several years, calculated according to the formula: $\text{CAGR} = (\text{ending value}/\text{starting value})^{1/\text{number of years}} - 1$.

[6] Huang X et al. Graphene-based composites. Chem Soc Rev 2012;41:666–86.

[7] The solid basis of these figures, gleaned from a wide variety of sources, is almost non-existent, amounting in most cases to a crude extrapolation of the trends of the past few years.

[8] Typically the entire carbon black production is counted as part of nanotechnology, but the size distribution of the particles in the material is very broad and only a minor fraction is smaller than 100 nm.

[9] The numbers given in this and the following paragraphs represent a considered consensus among a great variety of publicly accessible sources, too numerous to list individually.

[10] The volume of applications related to graphene is growing very rapidly. It is much more interesting than carbon nanotubes (CNT) as a material for active electronic devices. It is also superior for composites, because it has double the interfacial area per atom compared with CNT.

[11] Harris J, Ure D. Exploring whether 'nano-' is always necessary. Nanotechnol Perceptions 2006;2:173–87.

[12] *Project on Emerging Nanotechnologies: Consumer Products Inventory*. Washington (DC): Woodrow Wilson International Center for Scholars (the project began in March 2006).

[13] Berube DM. The magic of nano. Nanotechnol Perceptions 2006;2:249–55.

[14] Revell PA. The biological effects of nanoparticles. Nanotechnol Perceptions 2006;2: 283–98.

[15] Tran CL et al. A scoping study to identify hazard data needs for addressing the risks presented by nanoparticles and nanotubes. London: Institute of Occupational Medicine; 2005.

[16] This is very clearly not the case with highly elongated particles such as blue asbestos fibers, however; parallels with long carbon nanotubes are already raising concerns.

[17] It is perhaps a little surprising, given the weighty expertise that went into this report, that the outcome—the recommendations—is almost trivial. For example, it is considered that the top priority is the formation of a panel of well-characterized, standardized nanoparticles for comparison of data between different projects and laboratories, and the development of short-term *in vitro* tests aimed at allowing toxicity to be predicted from the physicochemical characteristics of the particles is recommended. Why, one may ask, was this not done before? Many scientists have worked on these problems already; it might have been more appropriate to examine why such a poor standard of experimentation has been accepted with so little criticism for so long.

[18] It is actually quite inadequate to refer generically to nanoparticles. It is already known that their toxicity depends on size, shape, and chemical constitution, and very possibly on their state of crystallinity (think of the problems of polymorphism of active ingredients in the pharmaceutical industry!). Therefore, at the very least some information on these characteristics should be provided when referring to nanoparticles.

[19] van Oss CJ et al. Impact of different asbestos species and other mineral particles on pulmonary pathogenesis. Clays Clay Minerals 1999;47:697–707.

Further reading

[1] Hunt G, Riediker M. Building expert consensus on problems of uncertainty and complexity in nanomaterials safety. Nanotechnol Perceptions 2011;7:82–98.

CHAPTER

Miscellaneous Applications

CHAPTER OUTLINE HEAD

6.1 Noncarbon Materials . . . 62
6.1.1 Composites . . . 62
6.1.2 Coatings . . . 63
6.2 Carbon-Based Materials . . . 63
6.3 Ultraprecision Engineering . . . 65
6.4 Aerospace and Automotive Industries . . . 66
6.5 Architecture and Construction . . . 66
6.6 Catalysis . . . 67
6.7 Environment . . . 67
6.8 Food . . . 69
6.8.1 Packaging . . . 69
6.8.2 Farming . . . 70
6.8.3 Sensors . . . 70
6.8.4 Nano-Additives . . . 71
6.8.5 Consumer Choice . . . 72
6.8.6 The Social Context . . . 72
6.8.7 Nanotechnology and the Food Crisis . . . 73
6.8.8 Conclusions . . . 73
6.9 Lubricants . . . 74
6.10 Metrology—Instrumentation . . . 75
6.11 Minerals and Metal Extraction . . . 75
6.12 Paper . . . 76
6.13 Security . . . 77
6.14 Textiles . . . 77

The remaining chapters of Part II survey commercial and near-commercial applications. Because of their importance, energy, information technology, and health applications are placed in separate chapters; all remaining applications are covered here. The order of coverage is firstly upstream technologies (i.e., those that enable a

Applied Nanotechnology, Second Edition. http://dx.doi.org/10.1016/B978-1-4557-3189-3.00006-3

wide range of applications), notably materials, carbon-based materials, and ultraprecision engineering, followed by downstream technologies in alphabetical order.

6.1 Noncarbon materials

The main raw nanomaterials manufactured on a large scale are nanoparticles. As already pointed out in Chapter 5, the bulk of them are traditional particles, notably carbon black, silver halide emulsion crystals, and pigments, which are in no sense engineered with atomic precision; some of them merely happen to fall within the accepted range of nano-objects [1]. These products are, in fact, typically quite polydisperse. Attempts to synthesize monodisperse populations on a rational basis have a considerable history [2]. Natural materials such as clays provide a significant source of nanoplatelets. Key parameters of nanoparticles are size (and size distribution), chemical composition, shape, and porosity.

6.1.1 Composites

Nanoparticles have relatively few direct uses; mostly their applications are in composites (i.e., a mixture of component A added to a matrix of component B, the latter usually being the majority component)—a nanocomposite differs from a conventional composite only insofar as the additive is nanosized and better dispersed in the matrix. The purpose of adding materials to polymer matrix is to enhance properties such as stiffness, heat resistance, fire resistance, electrical conductivity, gas permeability, and so forth; the object of any composite is to achieve an advantageous combination of properties. If the matrix is a metal, then we have a metal–matrix composite (MMC). A landmark was Toyota's demonstration that the incorporation of a few weight percent of a nanosized clay into a polyamide matrix greatly improved the thermal, mechanical, and gas permeability (barrier) properties of the polymer [3].

There is no general theory suggesting that the advantage scales inversely with additive size; whether a nanocomposite is commercially viable depends on all the parameters involved. There is such a huge variety of materials that it is perhaps futile to attempt a generalization. However, the very small size of individual nano-objects would make it feasible to incorporate a greater variety of materials within the matrix for a given total proportion of additive. Very often ensuring wetting of the nanoparticle—the most commonly used shape hitherto, although this will change as fragments of graphene, which is a nanoplate, increases in importance—by the matrix presents a significant technological hurdle. Most successful composites require the additive to be completely wetted by the matrix. Wetting behavior can be predicted using the Young-Dupré approach [4]; if, however, the particle becomes very small, the surface tension will exhibit a curvature-dependent deviation from the bulk value appropriate for a planar particle–matrix interface.

The chief manufacturing routes for polymeric nanocomposites are blending preformed nano-objects with the molten matrix; dispersing preformed nano-objects in the monomer precursor to the matrix and polymerizing it; and synthesizing the nano-objects *in situ* within the matrix. In all cases the composite may advantageously be prepared as a concentrated *masterbatch* (especially if effectively dispersing the nano-objects in the matrix requires some special technology) that is then shipped to the converter who can readily blend it with pure matrix polymer (e.g., if all the materials are pelleted, it may suffice to simply mix the pellets in the feed hopper of the converter).

6.1.2 Coatings

Many properties of materials essentially only concern their surfaces; if so, it is much more cost-effective to apply an expensive material as a thin film coating the bulk object in order to achieve the desirable interfacial attribute(s) (e.g., a low coefficient of friction). Furthermore, desirable bulk properties are not compromised. Alternatively, the surface can be modified using a technique such as ion implantation [5].

6.2 Carbon-based materials

Fullerenes, carbon nanotubes, and a graphene are true children of the Nano Revolution: they did not exist as commercial commodities before. Although carbon black and diamond-like carbon thin films have some nanofeatures, they are not atomically engineered and, moreover, existed before the era of nanotechnology; we do not propose to recruit them retrospectively. Figure 6.1 shows the striking growth (as measured by the annual number of publications) of carbon nanotubes, with a hint of saturation and the concomitant rise of graphene.

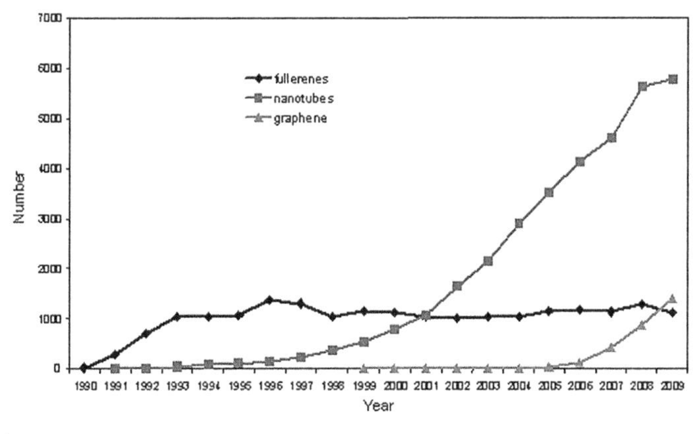

FIGURE 6.1

Annual world numbers of publications dealing with carbon nanomaterials.

Reproduced from A.I. Terekhov, Developing nanoresearch in Russia: a bibliometric evaluation. Nanotechnol. Perceptions 7 (2011) 188–198 with permission of Collegium Basilea.

Industrial problems associated with large-scale fullerene manufacture have been solved [6]; to date, their applications remain niche, however. Far more interest is attached to carbon nanotubes (CNTs) [7], and even more to graphene [8]. Some of the extraordinary properties of CNTs include: a very high aspect ratio (their diameter can be less than 1 nm, but they can be many micrometers long), making them suitable for field emission applications and as conducting additives to polymer matrices with a very low percolation threshold; very high electron mobility and the highest current density of any known material, approximately 10^9 A cm^{-2} (cf. copper with 10^6 A cm^{-2}, and aluminum 10 times less); ballistic electron transport; the highest Young's modulus of any known material (approximately 1 TPa along the axis); and the highest thermal conductivity of any known material (approximately 4000 W m^{-1} K^{-1}). Manufacturing problems seem to be on the way to being solved (note that some key applications use only very small quantities); the main issue today is probably purity. Nanotubes grown from a hydrocarbon feedstock such as acetylene using chemical vapor deposition require a metal catalyst (usually iron or nickel), which can be troublesome to remove afterwards; and preparations are frequently contaminated with amorphous carbon. The electrical properties will be revisited in Chapter 7; Table 6.1 summarizes the mechanical and thermal properties of carbon materials.

Because of their great strength, CNTs are especially attractive nanocandidates for incorporation into a matrix. Even more attractive than CNT are graphene fragments (nanoplatelets). The graphene is typically corrugated, due to numerous defects introduced during preparation, which is excellent for locking the material into the polymer (whereas CNT surfaces are usually smooth and defect-free). Per unit weight, the graphene anyway gives double the interfacial area compared with CNTs (which are, essentially, rolled up graphene). An additional advantage is that small sheets are much better at intercepting cracks in the composite than small tubes. This kind of composite effectively overcomes the reverse Hall–Petch relationship that prevents grain size from being usefully decreased down to the nanoscale in order to strengthen a material.

Their good field emission characteristics (including ultrahigh brightness and small energy dispersion) make CNTs outstandingly good electron sources for scanning electron microscopy—although the world market is very small. They are also useful in residual applications of vacuum tubes (e.g., high-power microwave amplifiers).

Table 6.1 Some properties of bulk and nanoscale carbon materials.[a]

Property	Unit	Diamond	Graphite	CNT
Young's modulus	N m^{-2}	10^9	10^{10}	10^{12}
Thermal conductivity	W m^{-1} K^{-1}	2000	20	3000

[a] *The given values are only approximate, in order to enable a rough comparative idea of the material properties to be formed. Actual measured values still depend on many experimental details.*

There is obviously an enormous potential application as field emission displays, although parallel innovations are required here, including a convenient means of positioning them. Furthermore, there is intense competition from organic light-emitting devices. The electronic properties (very high current densities) make CNTs attractive for the vertical wires (VIAs) connecting layers of integrated circuits, although exactly how their fabrication will be integrated into existing semiconductor processing technology has not been worked out. They may also be used as gates in field effect transistors (note that they can be prepared as semiconductors or metals). This configuration can be used to construct sensors for substances, if the gate is modified such that it interacts with the substance to be sensed. An even simpler configuration is merely to measure the longitudinal electrical resistance of the substance-sensitized nanotube.

The very low percolation threshold of these extremely elongated objects (a few volume percent or less) enables the preparation of electrically conducting polymers with such a low volume fraction of CNTs that the composite is visually unaffected by their presence. The main product of Hyperion (Chapter 13) is conducting paint suitable for the mass-production lines of the automotive industry. Other applications include antistatic coatings and electromagnetic screening films. Of great interest is the possibility of preparing transparent conducting films that could be used as the counterelectrode in displays. At present, indium-doped tin oxide (ITO) is the main material used for this purpose, but not only is it too brittle to be usable in flexible displays, but also the world supply of indium is expected to be exhausted within 2–3 years at present rates of consumption.

CNTs can also be used as the charge storage material in supercapacitors. All the atoms of single-walled nanotubes are on the surface, hence they have the highest possible specific surface area ($1.5 \times 10^3\ m^2\ g^{-1}$), suggesting a theoretical upper limit for energy density of $20\ W\ h\ kg^{-1}$. Supercapacitors rated at 1000 F are commercially available. Nevertheless, carbon black, which is much cheaper, is almost as good, hence it is doubtful whether this application will be commercially viable.

The very small size of a single nanotube makes it an attractive electrode material in electrochemical applications for which microelectrodes have already been shown to diminish transport-related overpotentials.

As far as composites are concerned, despite the extraordinarily high Young's modulus of CNTs, mechanical performance has not been demonstrated to be superior to that already attainable with carbon fibers. The problem is how to disperse the CNTs in the matrix.

The variable valence and the availability of electrons make CNTs attractive potential catalysts for certain reactions, for example in the petrochemical industry.

6.3 Ultraprecision engineering

The market for ultraprecision machine tools is relatively small, amounting to a few tens of millions of dollars annually. The USA has just two companies in this business, each one selling a few dozen machines a year; the machines themselves cost hundreds of thousands to a million dollars apiece.

6.4 Aerospace and automotive industries

The dominant goal is to reduce vehicle weight without compromising other relevant attributes. For spacecraft, launch is one of the highest cost factors and is directly related to weight, but aircraft and even road vehicles also benefit from reduced mass—less fuel is required to accelerate them. Hence there is much activity in seeking to replace the heavy metals used in components by lightweight polymers strengthened by nanoparticulate or nanofibrous additives (see Section 6.1.1). Other more specific aims include formulating lightweight electrically conducting materials for use in fuel lines to avoid the buildup of static electricity, ultrahard (abrasion-resistant) paint, low-friction finishes, and so forth. A significant difference between aerospace and automotive is that the lead time for introducing an innovation is typically 10 times longer in the former than in the latter (in which it is about 3 years), due to the more stringent needs for testing. Since sports equipment has many similar requirements, but is not usually safety-critical, it offers an interesting path for materials development and innovation to manufacturers.

6.5 Architecture and construction

The main application for nanotechnology in this sector is currently in materials, especially concrete enhanced using nanoparticles. Even though superior properties can be demonstrated, however, market penetration of nano-innovations can be expected to be slow, because of the traditional low-tech attitudes prevailing in much of the industry.

The continuing penchant of architects for designing large buildings predominantly covered in glass has provided a welcome countertendency, however, because glass offers many possibilities for nanotechnological enhancement. In particular, nanostructured superhydrophobic surfaces imitating those of the leaves of plants such as the lotus enable raindrops to scavenge dirt and keep the surfaces clean. Nanoparticles of wide band-gap semiconductors such as titanium dioxide can be incorporated into the surface of the glass, where they absorb ultraviolet light, generating highly oxidizing or reducing or both species able to decompose pollutants adsorbed from the atmosphere. Ultrathin film coatings, even of metals, can be applied to the surface of glass, in order to control light and heat transmittance and reflectance. Sophisticated glasses with electrically switchable transmittance are now available. "Anti-graffiti" paint, from which conventional paint sprayed on can be easily removed, has also gained a certain popularity (although a social, rather than a technological, solution might be more effective at eliminating unwanted graffiti).

Ultimately, the availability of dramatically new nano-engineered materials (e.g., ultrastrong and ultralight diamondoid panels) may usher in a totally new era of architecture.

6.6 Catalysis

It has long been recognized that the specific activity of heterogeneous catalysts increases with increasing state of division. This is, of course, an old market that has long been a very significant part of the chemical industry. World turnover amounts to almost $\$30 \times 10^9$ annually, a very large proportion of which could be supplied by nanotechnology. However, even though many catalysts use nanosize metal clusters (for example), they cannot be called examples of atomically precise engineering. Indeed, the whole field is remarkable for the high degree of empirical knowledge prevailing. In the future, nanotechnology offers the chance to assemble catalysts atom-by-atom. There is a general feeling in the industry that there is still considerable potential for increasing the activity of catalysts (through both more effective acceleration of the desired reaction and more effective suppression of undesired side reactions).

About a quarter of the world market is accounted for by oil refining, and over half is nowadays accounted for by automotive exhaust catalysts.

6.7 Environment

"Environment" is a concept even more amorphous than energy in a commercial context. Here, we mainly consider the remediation of contaminated soils and groundwater by the addition of nanoparticles [9]. Regarding the former, if a source of ultraviolet light is available (sunlight is adequate), titanium dioxide is a useful material; absorption of light creates electron-hole pairs acting as strong reducing–oxidizing agents for a large variety of organic compounds adsorbed on the nanoparticle surface. By this means many recalcitrant potential pollutants can be destroyed [10]. Until now attention has been mainly concentrated on the actual science of the photoassisted chemical decomposition rather than devising a complete process in which the nanoparticles, having done their work, would be collected and possibly regenerated for further use. A very convenient way to accomplish this is to first produce superparamagnetic nanoparticles (e.g., from magnetite) and then to coat them with the photocatalytic material to make a "core–shell" particle that can be conveniently displaced using an external magnetic field.

Soil remediation is also mainly concerned with eliminating pollution. In particular, iron-containing nanoparticles are being promulgated as superior alternatives to existing remediation procedures using more or less comminuted scrap iron for soil contaminated with chlorinated hydrocarbons—their decomposition is catalyzed by magnetite (Fe_3O_4). Hence the addition of nanoparticulate iron oxide to soil is a possible remediation method. Unfortunately, there is minimal documented experience to guide the would-be practitioner. These environmental applications would have to operate on a large scale in order to be effective; the effects of releasing large numbers of nanoparticles into the biosphere are not known. Iron is generally presumed to be a rather benign element, but nanoparticles may be able to penetrate into the microbial organisms ubiquitous in soils with unknown effects on their vitality and on interspecies interactions. In other words, this proposed technology raises a number of questions—to start with there does not seem to be any conclusive evidence that it is actually efficacious, and furthermore there is the still open question of the effect of dispersing a significant concentration of nanoparticles (and it has to be significant, otherwise there would be no significant remediation) on the ecosystem, especially on microbial life in the soil.

One often hears it stated that nanotechnology will enable the environment to be returned to a pristine state, without an accompanying explanation of the process by which this rather vague assertion might be realized. It seems that there are going to be two principal impacts of nanotechnology on the environment. The first is immediate and direct (the use of nanoparticles for cleaning up pollution, as just discussed), the second long term and indirect. The latter effects follow from the obvious corollary of atomically precise technologies—they essentially eliminate waste during fabrication. This applies not only to the actual fabrication of artifacts for human use, but also to the extraction of chemical elements from the geosphere (should those elements still be necessary). If the manufacture of almost everything becomes localized, the transport of goods (a major contributor to energy consumption and environmental degradation) should dwindle to practically nothing; the localized fabrication implied by the widespread deployment of productive nanosystems and personal nanofactories should eliminate almost all of the currently vast land, sea, and air traffic involved in wholesale and retail distribution; this elimination (and a commensurate downscaling of transport infrastructure) will doubtless bring about by far the greatest benefit to the environment of any aspect of nanotechnology. Furthermore, atom-by-atom assembly of artifacts implies that discarded ones can be disassembled according to a similar principle, hence the problem of waste (and concomitant environmental pollution) associated with discarded obsolete objects vanishes.

In the intervening period, the general effect of nanotechnology in promoting energy efficiency (see Chapter 7) will of course be beneficial to the environment—less transport of fossil fuels, less pollution from combustion, and so forth.

Concerns have, however, been expressed that some of the nanoparticles being promoted for use in a variety of products, especially those with some kind of antibiotic activity, among which silver nanoparticles are the most widespread, are harmful if released into the environment. The environment contains innumerable bacteria and other microbes, and nanoparticles designed for their antibacterial activity are evidently going to continue that activity wherever they are. Release is almost inevitable. Many textiles are already sold conjugated with silver nanoparticles; some of them will be released when the textiles are laundered. Once nanoparticles, not just silver ones, and other nano-objects become routinely produced industrially, it is inevitable that they will find their way into watercourses and soils. Food chains will ensure that they are disseminated throughout the ecosystem. Being ultrasmall, nanoparticles are highly mobile and it will be difficult to absolutely contain them during the manufacture of finished products incorporating them. Furthermore, there can be little control over their fate once the artifact containing them is discarded.

This is why there are already widespread calls for stricter regulation of the deployment of nanotechnology, particularly nano-objects in consumer products. These calls are driven by a growing awareness of the potential dangers of nano-objects penetrating into the human body, and by the realization that understanding of this process is still rather imperfect, and prediction of the likely effects of a new kind of nano-object is still rather unreliable. Furthermore, there have been a sufficient number of cases, albeit individually on a relatively small scale, of apparently unscrupulous entrepreneurs promoting nano-object-containing products that have turned out to be quite harmful.

These calls, while seemingly reasonable, raise a number of difficulties. One is purely practical: once nano-objects are incorporated into a product they are extraordinarily difficult to trace. Traceability is only feasible up to the point of manufacture, and even then only if the manufacturer has sourced materials through a regulated or self-regulated commercial channel such as a commodity exchange. Establishing the provenance of nanoparticles that might turn up in a waste dump, for example, poses a very difficult forensic challenge.

Furthermore, regulation will become essentially meaningless if productive nanosystems become established: every individual would be producing his or her own artifacts according to his or her own designs and it is hard to see how this could be regulated in a free society.

Hence, we conclude that although on the whole nanotechnology must have a favorable impact on the environment, there are also some problematical aspects, which might be partly alleviated by regulation (cf. Chapter 12).

6.8 Food

6.8.1 Packaging

A useful application of nanotechnology in the food industry is to enhance packaging. A major problem of industrial food preparation and dissemination is keeping the products fresh and fit for human consumption. If packaging can more effectively act as a barrier (e.g., to oxygen), food can be kept fresher for longer prior to sale and opening the package. The best flexible packaging is a film of aluminum, sandwiched between polymer films, which has essentially zero permeability but is opaque, whereas it would be desirable for the package to be both impermeable and transparent. The functional enhancement of an aluminum-free film is essentially achieved by transforming the polymer into a nanocomposite (see Section 6.1.1), by adding nanoplatelets (e.g., exfoliated clays), which enormously increase the tortuosity of diffusion pathways through the film. Worldwide sales of nanotechnology products to the food and beverage packaging sector have been rapidly increasing—$150 million (2002), $900 million (2004), $4100 million (2008), and $7300 million (forecast for 2014). There were less than 40 identifiable nanopackaging products in the market 7 years ago, whereas there are more than 250 now.

Major market trends in the food and beverage sector include: improving the performance of packaging materials in a passive sense (e.g., by increasing their transparency); prolonging the shelf life of the contents (e.g., by selectively managing the gas permeability of the packaging); improving sterility (e.g., by immobilizing antibiotics that kill microbes on contact within the packaging material); indicator packaging (which changes color if the package has been subjected to a deleteriously high temperature, for example, which renders the contents unfit for consumption but otherwise leaves no visible traces); interactive packaging (which might respond to a potential customer touching it by changing color—or, indeed, by emitting a yelp or a purr, as a ploy to increase sales); and tagged packaging (incorporating unique, multicomponent nanoparticles encoding the tagging information). In other words, nanotechnology allows molecular-scale structural alterations of packaging materials. With different nanostructures, polymer films can gain or lose various gas and water vapor permeabilities to fit the requirements of preserving fruit, vegetables, beverages, wines, and other foods. By adding nanoparticles, manufacturers can produce bottles and packages with more light and fire resistance, stronger mechanical and thermal performance, and less gas absorption [11]. Such nano-tweaking can significantly increase the shelf life of foods and preserve flavor and color. Nanostructured films coating the package can prevent microbial invasion and ensure food safety. With embedded nanosensors in the packaging, consumers will be able to determine whether food has gone bad or estimate its nutritional content. These functionalities are increasingly necessary in an urban environment, where food is anything but locally produced.

6.8.2 Farming

The main effect of nanotechnology is to introduce more precision into agriculture. Nano-enabled, powerful microprocessors enable computation to be all-pervasive. The farmer can integrate satellite and field data to optimize plowing, fertilization, and planting schedules and can even use a geographical information system to drive robots in his fields. Nanofertilizers can be custom-formulated for the particular conditions found in each of his fields, and the constituents of the fertilizers may themselves be nano-enabled, such as the nanochelates that are turning out to be especially useful for the delivery of essential metal ions. In the future, butchers may routinely employ tomography on carcasses to determine the optimal dissection. The tomography itself requires heavy computations; nanotechnology-enabled processing power may become powerful enough to enable the optimal dissection to be automatically executed by robots as well. Cold storage systems—and indeed the logistics of the entire global distribution system—are nowadays efficiently controlled by microprocessors.

6.8.3 Sensors

Apart from the sensorial materials used in packaging, nanoscale chemical sensors, cheap and unobtrusive enough to be ubiquitous, can be deployed to monitor the state of food, including possible contamination with pesticides, or infectious agents acquired in the factory, or deficiencies arising through improper operations in a restaurant kitchen. Driven by a plethora of scandals leading to food poisoning, sometimes on quite a large scale, this is perceived to be a very welcome development by the general public, significantly offsetting some of the disadvantages of the modern agro-industrial complex. A very recent development is nano-enabled microsensors able to swiftly determine the nutritional benefit of a food to the person considering consuming it (whose genetic and metabolic characteristics of the gastrointestinal tract have been previously assessed) [12].

A good example of how nanoscale experimentation using a kind of sensor has led to a profoundly new understanding of previously unsuspected hazards is provided by the discovery, using the black lipid membrane (BLM) technique [13], that certain cyclic polyunsaturated compounds synthesized by bacteria used for the biotechnological production of artificial sweeteners are able to form ion channels in our cell membranes. Trace quantities of these compounds remain in the so-called "high energy" and other soft drinks that seem to enjoy a growing popularity, and may be responsible for the growing incidence of heart problems among teenagers in societies where the consumption of these beverages has become the norm. Knowing this is one thing; it is another matter to diffuse the knowledge among the general public, in order that they may weigh the risks against the supposed enjoyment.

6.8.4 Nano-additives

The inclusion of nanoscale nutritional additives in food, another direct application, is, however, contentious [14]. Soluble or microscale additives are already widespread in the processed food industry (a very common example is the addition of vitamin C to fruit juices); the idea behind using nanotechnology is to enhance their functionality and hence effectiveness—for example, encapsulating the vitamin C in hollow nanospheres made from calcium carbonate, so that the vitamin does not oxidize and become nutritionally valueless while the juice is standing in the air waiting to be drunk, but will only be released in the strongly acidic environment of the stomach. Inasmuch as these additives are already becoming more and more sophisticated, introducing nanotechnology seems to be merely a continuation of an existing trend. More contentious are "neutraceuticals"—foodstuffs deliberately enhanced with substances that would rank as pharmaceuticals. An example is the addition of fluoride to drinking water for the sake of dental benefits, although not all people drinking the water require them (the controversy that such addition has attracted seems to be justified on the grounds that the benefits of fluoride are realized when it is applied topically to the teeth, not ingested). Nanotechnology could facilitate a vast expansion of such enhancements. The fundamental argument against this kind of thing is that our physiology is not adapted to such novelties and may not be able to adapt before some harm is done. (This constitutes the basic objection to ingesting genetically modified foods.) It is quite difficult to find the right level at which to address the problem. Clearly DNA as such is not in general toxic—we are eating it all the time. On the other hand, *certain sequences* (e.g., those of a virus) are demonstrably harmful, at least under certain circumstances. The situation recalls the debates over the quality of drinking water in London in the middle of the 19th century—certain experts likened the inadvertent consumption of microorganisms in the water supply as being no more dangerous than eating fish. Given the state of knowledge at the time, it was not easy to refute that argument.

6.8.5 Consumer choice

An important aspect of the current debate on the matter concerns the possibility of choice. Ideally, if knowledge is insufficient for it to be clear whether benefits or risks are preponderant, a product should be available both with and without the nano-additive. Then the consumer can make his or her choice—*caveat emptor* [15]. In reality, choice tends to contract. For example, nearly all the world's soybeans come from a certain (genetically modified) variety; 99% of tomatoes grown in Turkey are no longer indigenous varieties [16]. It appears to be empirically well established that mysterious "market forces" drive matters to this result, and the presence or absence of choice needs to be taken into account when it is discussed whether nano-additives should be permitted or not. We are familiar with the state of affairs in traditional (non-nano) food processing. For example, it is possible to buy dairy products, such as cheese, made from either raw or pasteurized milk. For some consumers, avoiding the risk of contracting tuberculosis is the preponderant consideration; for others, the undesirability of consuming advanced glycation end-products (AGEs, resulting from the chemical reaction between sugars and animal proteins or fats, typically taking place during frying or pasteurization) outweighs that risk; for yet others taste is the most important consideration.

6.8.6 The social context

"Man ist, was man isst," as Martin Luther famously remarked in one of his *Tischgespräche*. Given the centrality of food for human existence, it can hardly be discussed in isolation as a purely physiological matter. Indeed, there is even evidence that the intake of folic acid by a pregnant mother can influence the methylation of the baby's proteins [17]. Furthermore, it is perhaps too much to expect that we always eat "sensible" foods, or carefully examine the list of ingredients on a packet (which anyway is usually woefully inadequate, particularly regarding the actual quantities of the substances mentioned), or acquire a personal biosensor for verifying on the spot the absence of hormone-active substances in vegetables. It may be nowadays trite to repeat John Donne's dictum "No man is an island," but if anything it is even more true today, in a world wide web-connected age, than it was at the end of the Middle Ages. We are all affected by the foods around us, whether we partake of them or not.

The dominant social aspect of nutrition is malnutrition coupled with obesity. It is estimated that current world production of food is adequate for the current world population, but much of that food is in the wrong place at the wrong time. Technologies, such as cold storage and nanoparticle-containing gas-resistant wrappers, should, therefore, contribute to alleviating unevennesses of supply. The technologies come at a price, however—for example, many modern farming practices, including intensive agriculture of all kinds and fish farming, tend to yield produce that is less wholesome than their nonintensive counterparts (especially regarding their micronutrient content).

The solution to eating the wrong foods—such as those that leave one undernourished or overweight or both—is surely more knowledge. This is the perennial problem of a society based on high technology—it can only be truly successful if all members are sufficiently knowledgeable to properly partake in its development. Hence, we also need to inquire how nanotechnology can contribute to the education of the population [18].

6.8.7 Nanotechnology and the food crisis

In June 2008 the Food and Agriculture Organization (FAO, part of the United Nations) held a 3-day summit conference in Rome in order to explore ways of overcoming problems caused by steeply rising food prices around the world, which have caused especially grave problems in poor countries. Although part of the problem lies in the commercial sphere, and may be dealt with by considering the effects of export restrictions and price controls, a sustainable solution clearly lies in the technological realm. In the short term, charitable deliveries of seeds and fertilizers may alleviate the problem; in the medium term such measures are likely to make things worse. Therefore, a thorough appraisal of the state of agronomy is needed. In fact, a number of recent reports have pointed to the research deficit in the field accumulated during the last few decades [19]. Yet in its call for increasing public support (in the developing countries) for agronomy research, the FAO is essentially still thinking of traditional approaches. In view of the generally revolutionary nature of nanotechnology, it must be expected that here too it can make a decisive contribution, by the intense nanoscale scrutiny of the processes of comestible biomass production in its entirety, followed by inspired intervention at that scale. An example is biological nitrogen fixation. A wealth of detail is already known about the process: at the molecular level (the nitrogenase enzymes responsible for actually fixing the nitrogen); at the microbiological level (the symbiotic rhizobia); and at the ecological level (soil and inoculation). Intervention, e.g., with functionalized nanoparticles, especially to improve nitrogen fixation in difficult (e.g., dry or saline) conditions, seems to be feasible. The actual need is for laboratory and field research to establish the possibilities and limitations of such an approach.

6.8.8 Conclusions

Looking back over the past millennia of human civilization, improvements in technology have enormously increased agricultural output, but this has also led to a concomitant increase in world population, hence global nutritional difficulties remain. Geographical mismatch of supply and demand is frequently mentioned as a contributor to malnutrition—somewhat ironically, in an age of unprecedentedly large global trade volumes. A serious current problem is that it is becoming increasingly clear that productivity increases imply product quality decreases. This goes well beyond mere unpalatability [20]. The output of the agro-industrial complex, unfortunately including residues of pesticides and the presence of hormone-active substances, may solve the basic malnutrition problem, but may introduce new problems of ill health that seem to be deeply unsustainable, although the manufacturers of pharmaceuticals may see it as a source of new opportunities.

6.9 Lubricants

Low-friction coatings are classically made from layered materials such as graphite (C), molybdenum disulfide (MoS_2), or tungsten disulfide (WS_2). As well as the transition metal dichalcogenides, bismuth selenide (Bi_2Se_3) and telluride and boron nitride (BN) are also important, and some transition metal oxides (e.g., MoO_3) [21]. Physical vapor deposition (e.g., sputtering) is suitable for applying the coatings. Coating thicknesses are typically several micrometers and are not therefore in the nanoscale. The layered sulfides can yield a friction coefficient of less than 0.05 (0.005 appears to be possible), depending on environmental conditions (including humidity). This compares with steel on steel, which has a friction coefficient of about 0.8, falling to about 0.1 when lubricated with motor oil. Wear is also an issue, which is why thicknesses of less than a few hundred nanometers are not recommended. Some of these materials do not perform well over a wide range of temperatures, especially in air, in which oxidation of the non-oxide materials might become problematical at higher temperatures.

Composite films can be used to combine properties. For example, cosputtering gold and molybdenum sulfide creates a low-friction electrically conducting coating useful for sliding electrical contacts [22]. More sophisticated composites are designed to adapt their structure and surface composition in order to maintain their tribological properties under varying conditions. For example, silver and molybdenum nanoparticles in an yttrium-stabilized zirconium oxide matrix, produced by cosputtering, provide a silver-rich surface for low friction at temperatures up to 500 °C, above which diffusion causes the surface to become enriched in MoO_3; the diffusion can be controlled by incorporating a porous TiN layer within the structure [23]. The long-term behavior of these materials under storage is unknown; at low temperatures (i.e., after fabrication) the structure is metastable.

A successful low-friction coating might make the use of a liquid lubricant superfluous (which is anyway likely to be problematical in electrical applications).

As an alternative to precoating surfaces, the materials can be added to the lubricant in the form of nano-objects (e.g., nanoplatelets, which can be made from a wide variety of tribologically advantageous materials [24] or nanocopper) to lubricating fluids. The observed effects, which may include wear reduction, depend on the deposition of the nano-objects on the moving surfaces.

It is typical of the intermediate situation in which we currently find ourselves that nanotechnology can sometimes be used to enhance "subnano" technologies (i.e., finished with a precision lower than that of nanotechnology), the need for which will ultimately disappear. For example, if machining can routinely achieve a surface roughness of 1 nm combined with ultrahardness, lubricants may not be necessary, but as long as surfaces are still rougher or softer, nano-engineered lubricants (e.g., Au/MoS_2 solid lubricant films) offer better performance than conventional lubricants.

Other examples of such intermediate or bridging technologies are to be found in the field of thermal management (Section 8.2).

A very hard carbon film called "near-frictionless carbon" (NFC), better than diamond-like carbon, has been developed by the Argonne National Laboratory. It combines low friction (achieving a friction coefficient of about 0.04 in air and as little as 0.001 in dry argon) with great hardness. Hence, wear rates are extremely low ($< 10^{-10}\ mm^3\ N^{-1}\ m^{-1}$, compared with $10^{-7}\ mm^3\ N^{-1}\ m^{-1}$ for unlubricated steel and $10^{-7}\ mm^3\ N^{-1}\ m^{-1}$ for conventionally lubricated steel).

6.10 Metrology—instrumentation

The primary products included in this category are the scanning probe microscopes that are indispensable for observing processes at the nanoscale, and which may even be used to assemble prototypes. The market is, however, still relatively small in value—estimated at around $250 million per annum for the whole world. This represents about a quarter of the total microscope market. Optical microscopes have similar share (but are presently declining, whereas scanning probe microscopes are increasing), and electron microscopes have half the global market.

Given the growing importance of metrology for the rational development of nanoproducts (see Section 10.9), this sector must inevitably grow and become more diverse. While the first instruments of a new type are built in the laboratory of the inventors, if it becomes apparent that there is widespread demand for them, commercial instrument manufacturers quickly move to provide user-friendly exemplars (this has been a very striking development with scanning tunneling and atomic force microscopes, both of which were typically constructed in the laboratories of institutions wishing to use them in the years immediately following their invention).

At the other end of the size spectrum are telescopes, looking at very large and very distant objects. New generations of space telescopes and terrestrial telescopes used for astronomy require optical components (especially mirrors) finished to nanoscale precision. The current concept for very large terrestrial telescopes is to segment the mirrors into a large number of unique pieces (of the order of 1 m^2 in area), the surface of each of which must be finished to nanoscale precision [25].

6.11 Minerals and metal extraction

The current technologies based on pyrometallurgy used to extract metals from ores use vastly more energy than is theoretically required [26]. For example, currently a sulfide ore might be roasted to create the oxide, which is then reduced using carbon monoxide at high temperature. Not only is much energy required to maintain the high temperature of the process, but free energy is also wasted in the sense that the reaction of metal sulfides with oxygen is energetically downhill. There is also an enormous production of waste material (e.g., sulfur oxides), the disposal of which is damaging to the environment.

Nanotechnology can be brought to bear on this problem in many different ways. Biomimicry seems a very attractive route to explore, especially since living organisms are extremely good at extracting very dilute raw materials from their environment, operating at room temperature and, seemingly, close to the thermodynamic limit (i.e., in a highly energetically efficient manner). The kidney is perhaps the most remarkable example of this nature [27]; diatoms have the remarkable ability to concentrate silicon from the sea for use in their rigid cell walls. Nano-engineered "artificial kidneys" could be used not only to extract desirable elements from very dilute sources, such as seawater, but also to extract elements or extractable compounds from natural waters polluted with them. Catalysts will, in this field too, doubtless play a very important role, especially if nanotechnologists succeed in creating durable artificial enzymes able to catalyze the necessary reactions at room temperature.

In order to directly harness this general ability of living organisms to extract metals, higher forms of life are less suitable than prokaryotic microbes (bacteria and archaea), because prokaryotes have far more diverse metabolisms than eukaryotes. In prokaryotes it seems that for every naturally occurring metal there is a gene for handling it! Lithotrophic prokaryotes are perhaps of the greatest interest, especially those converting sulfides to soluble sulfates (and using this reaction as their energy source). They are indeed presently used in a few dozen plants around the world to leach cobalt, copper, gold, nickel, and zinc from their ores. Organisms expressing metal reductases may have even greater potential for extraction. Another approach is to use complex biomolecules (proteins) to specifically bind the metals to be extracted. Microbes proliferate exponentially and hence essentially unlimited quantities can be rapidly and usually economically obtained; the same applies to the products of microbial metabolism, such as metal-binding proteins. Furthermore, one is no longer restricted to what nature has evolved, but specific functionality can be engineered into a molecule or, indeed, into the entire microorganism.

6.12 Paper

This commodity is made in vast quantities (globally, about 100 million tonnes per annum) in most countries in the world. The primary constituent is cellulose fiber, but as much as half of the annual production of paper contains small particles (typically 10–20% of the total mass). The use of such fillers in papermaking has a long history [28]. The purpose is partly economic, the fillers being generally cheaper than the cellulose, and partly to better control attributes such as porosity, reflectance, ink absorption (printability), permeability, stiffness, and gloss. The particles may be present as a surface coating or incorporated within the bulk of the sheet. A new application for nanoparticles is to tag sheets of paper with distinguishable nanoparticles (for example, made up from different metals) for security and identification purposes. Such particles would, of course, only need to be present in minute quantities. Individual cellulose fibers are being coated with nanoscale polyelectrolyte films in order to enhance strength and other attributes of paper such as electrical conductivity [29].

The coating is a self-assembly process whereby the fiber is merely dipped in a solution of the polyelectrolyte. Ease of manufacture makes the treatment quite cost-effective (e.g., by doubling the tensile strength, single-ply sacks can be used instead of double-ply, but the cost per unit area of the paper is less than double that of the untreated material).

6.13 Security

Military history is peppered with examples of lightly armed armies besting heavily armed opponents, by virtue of the superior mobility and speed of response conferred by the light weaponry. English schoolchildren know that the battle of Agincourt (1415) was won by the English because of the decisive superiority of their light longbows, an innovative, superior technology in comparison with the powerful but heavy crossbow favored by the French. Even the much earlier (ca. 1000 BC) story of the lightly armed, hence agile, David beating the lumbering Goliath illustrates this general principle at the level of the individual warrior; the innovation of the battlecruiser, championed by Admiral Fisher at the beginning of the 20th century, versus the heavier, slower battleship may be considered as yet another example. Seen in this light, nanotechnology is simply the latest means whereby military ordnance can be reduced in size and weight while keeping the same performance or even enhancing it.

Although military organizations such as the Department of Defense in the USA are spending a great deal on nanotechnology, most of the applications are generic, covering almost every aspect, ranging from essentially civilian applications such as ultralight clothing and footwear for an individual warrior to the sophisticated information-processing electronics that can be embedded in almost everything. In other words, most military applications of nanotechnology currently under investigation are adaptations of civilian products (and many can be sourced directly from the civilian market) [30].

Homeland security is heavily focused on the detection of explosives. This calls for chemical sensors of trace volatile components, using the same kind of technology as is used for medical diagnostics applications (see Chapter 9). Nanotechnology also enters into the video surveillance technology rapidly becoming ubiquitous in the civilian world, notably through the great processing power required for automated pattern recognition.

6.14 Textiles

A natural textile fiber such as cotton has intricate nanostructure; the comfortable (to feel and to wear) properties of many traditional textiles result from a favorable combination of chemistry and morphology. Understanding these factors allow the properties of natural textiles to be equaled or even surpassed by synthetic materials. Furthermore, nano-additives can enhance textile fibers with properties unattainable in the natural world, such as ultrastrength, ultradurability, flame resistance, self-cleaning capability, modifiable color, antiseptic action, and so forth. Textiles releasing useful chemicals, either passively or actively, are also conceivable (of which the antiseptic textile, in which silver nanoparticles are incorporated, is a simple example; such functionally enhanced textiles are typically used in specialty applications, such as serving as a living cell scaffold assisting tissue regeneration, and as wound dressings; it should not be forgotten that these nanoparticles are fairly fugitive and may end up polluting the environment—see Section 6.7).

The addition of nano-objects to textiles essentially creates a particular kind of nanocomposite. It is also possible to nanostructure the textile fiber surface and to weave the textile from a nanoporous material. The nanomodification can be carried out either during the preparation of the fibers or during a subsequent stage, working with the finished fibers or even with the finished woven fabric. In general, nano-objects embedded within the fiber rather than on the surface using covalent chemical bonds will provide the most durable composition. Table 6.2 gives an indication of the relationship between nano-object type and novel property conferred onto the textile.

A general-purpose technology for fabricating nanofibers is electrospinning: a high electric field is applied to a liquid droplet (e.g., of a polymer solution), electrostatic repulsion opposes the surface tension to stretch the droplet until, above a certain threshold, a *Taylor cone* is formed [31], from which a stream of liquid emerges (if the molecular cohesion of the liquid is insufficient to prevent the stream breaking up, the phenomenon of electrospraying occurs); elongation and thinning of the fiber as the stream dries in-flight results in uniform nanofibers [32]. Inorganic nanofibers have been made by an ingenious extension of the process, in which an organometallic

Table 6.2 Nanoparticles conferring certain attributes onto textiles.[a]

	Nanoparticle or Other Nano-Object							
Attribute	**Ag**	**Al_2O_3**	**C Black**	**CNT**	**Clay[b]**	**SiO_2**	**TiO_2**	**ZnO**
Abrasion-resistant		×		×	×	×		×
Active carrier				×	×			
Antimicrobial	×					×	×	
Antistatic	×		×	×				
Chemical-resistant		×				×		
Dirt-repellent						×	×	×
Electrically conductive	×		×	×				
Flame-retardant		×		×	×	×	×	
Photocatalytic							×	×
Self-cleaning	×					×	×	×
Tear-resistant				×?				
UV-absorbing							×	×
Water-repellent						×	×	×

[a] *Source: Grundlagen und Leitprinzipien zur effizienten Entwicklung nachhaltiger Nanotextilien. St. Gallen: Empa and TVS Textilverband Schweiz (2011).*

[b] *Typically exfoliated montmorillonite.*

precursor is mixed together with the polymer, and the fibers are subsequently pyrolyzed [33]. Inorganic nanofibers can also be produced by electrospinning molten ceramics at high temperatures.

References and notes

[1] This remark does not do justice to the extraordinary sophistication of a fabricated photographic emulsion crystal, achieved during more than a century of intensive research. For a historical overview, see Hercock RJ, Jones GA. Silver by the ton. London: McGraw-Hill; 1977.

[2] Uniform grain size was an important goal of emulsion manufacturers (see [1]). For other semiconductors see, e.g., Ramsden JJ. The nucleation and growth of small CdS aggregates by chemical reaction. Surf Sci 1985;156:1027–39; and Graham T. On liquid diffusion applied to analysis. J Chem Soc 1862;15:216–69 for an example of much earlier work that achieved nanoparticle monodispersity.

[3] A much older composite is paint, which consists of a pigment (quite possibly made of nanoparticles) dispersed in a matrix of varnish. Paint combines the opacity of the pigment with the film-forming capability of the varnish. Another mineral–polymer composite is the material from which many natural seashells are constructed: platelets of aragonite dispersed in a protein matrix. In this case, however, the matrix only constitutes a few percent of the volume of the composite.

[4] For an introduction, see Cacace MG et al. The Hofmeister series: salt and solvent effects on interfacial phenomena. Quart Rev Biophys 1997;30:241–78.

[5] See Fraunhofer Gesellschaft. Produktionstechnik zur Erzeugung funktionaler Oberflächen. Status und Perspektiven Brunswick. 2008; and Ramsden JJ et al. The design and manufacture of biomedical surfaces. Ann CIRP 2007;56(2):687–711.

[6] Arikawa M. Fullerenes—an attractive nano carbon material and its production technology. Nanotechnol Perceptions 2006;2:114–21.

[7] Boscovic BO. Carbon nanotubes and nanofibres. Nanotechnol Perceptions 2007; 3:141–58.

[8] Huang X et al. Graphene-based composites. Chem Soc Rev 2012;41:666–86.

[9] Rickerby D, Morrison M. Prospects for environmental nanotechnologies. Nanotechnol Perceptions 2007;3:193–207.

[10] E.g., Hidaka H et al. Photoassisted dehalogenation and mineralization of chloro/fluoro-benzoic acid derivatives in aqueous media. J Photochem Photobiol A 2008; 197:115–23.

[11] Ramsden JJ. Nanotechnology in coatings, inks and adhesives. Leatherhead: Pira International; 2004

[12] Vergères G et al. The NutriChip project—translating technology into nutritional knowledge. Br J Nutrition 2012;108:762–8.

[13] Grigoriev PA. Unified carrier-channel model of ion transfer across lipid bilayer membranes. J Biol Phys Chem 2002;2:77–9.

[14] Less controversial than nano-additives is the deliberate nanostructuring of processed food in order to affect "mouthfeel". This is a very difficult characteristic of a natural foodstuff to imitate; hence the food processing industry is devoting a great deal of ingenuity to find out how to do it. Nanometrology (Section 10.9) is an essential part of this endeavor.

[15] This, incidentally, highlights the importance of members of society being sufficiently knowledgeable to be able to make an informed choice (see also Chapter 18).
[16] For further examples, see Ramsden JJ. Complex technology: a promoter of security and insecurity. In: Ramsden JJ, Kervalishvili PJ, editors. Complexity and security. Amsterdam: IOS Press; 2008. p. 249–64.
[17] The methylation pattern of the gene repertoire is a key controlling factor in development.
[18] *Caveat emptor* is a universal injunction, which should actually render regulation superfluous, since it is generally called for to protect the unsuspecting consumer from unscrupulous purveyors of goods or services, but if the consumer took the trouble to properly investigate what he was letting himself in for, presumably a good or service that was not what it purported to be would not be bought, and the unscrupulous purveyor would be less likely to continue to attempt to sell whatever it was, profit presumably being the sole motive. This is an example of how market forces should work for the benefit of society. Yet, despite the repeated assertion that developed countries are "knowledge-based economies," it seems that the knowledge necessary to properly apply the principle of *caveat emptor* is lacking among the general public—and in many countries is lacking among the legislative bodies. Perhaps the future availability of nano-enabled sensors for food, water, ambient air, etc., which may become as ubiquitous as mobile phones are now (indeed they could be incorporated into mobile phones) will provide the impetus for the general public to apply knowledge to select what is consumed.
[19] World Development Report 2008: Agriculture for development. Washington, DC: World Bank; 2007.
[20] "Mere" perhaps belies the significant contribution of the enjoyment of food to social harmony, creativity, etc.
[21] See Khorramian BA. Solid lubricant for low and high temperature applications, US Patent no 5,747,428; 1998 for a variety of other compounds.
[22] Lince JR. Tribology of co-sputtered nanocomposite Au/MoS_2 solid lubricant films over a wide contact stress range. Tribol Lett 2004;17:419–28.
[23] Muratore C et al. Adaptive nanocomposite coatings with a titanium nitride diffusion barrier mask for high-temperature tribological applications. Thin Solid Films 2007;550:3638–43.
[24] Coleman JN et al. Two-dimensional nanosheets produced by liquid exfoliation of layered materials. Science 2011;331:568–71.
[25] Shore P. Ultra precision surfaces. In: Proceedings of the ASPE, Portland, OR; 2008: 75–80.
[26] Gillett SL. Nanotechnology, resources, and pollution control. Nanotechnology 1996;7:177–82.
[27] Thomas SR. Modelling and simulation of the kidney. J Biol Phys Chem 2005;5:70–83.
[28] Hubbe MA. Emerging technologies in wet end chemistry. Leatherhead: Pira International; 2005.
[29] Zheng Z et al. Layer-by-layer nanocoating of lignocellulose fibers for enhanced paper properties. J Nanosci Nanotechnol 2006;6:324–32.
[30] Altmann J. Military nanotechnology. London: Routledge; 2006.
[31] Taylor G. Disintegration of water drops in an electric field. Proc Roy Soc A 1964;280:383–97.

[32] Li D, Xia Y. Electrospinning of nanofibers. Adv Mater 2004;16:1151–70; Pham QP. Electrospinning of polymeric nanofibers for tissue engineering applications. Tissue Eng 2006;12:1197–211.
[33] Chuangchote S et al. Photocatalytic activity for hydrogen evolution of electrospun TiO_2 nanofibers. Appl Mater Interfaces 2009;1:1140–3.

Further reading

[1] McKeown P et al. Ultraprecision machine tools—design principles and developments. Nanotechnol Perceptions 2008; 4: 5–14.
[2] Mitov M. Sensitive matter. Cambridge, MA: Harvard University Press; 2012.

CHAPTER

Energy

7

CHAPTER OUTLINE HEAD

7.1 Energy Harvesting . . . 84

7.2 Production and Storage . . . 84

7.2.1 Production . . . 85

7.2.2 Storage . . . 87

7.3 Energy Efficiency . . . 91

7.3.1 Lighting and Computation . . . 92

7.3.2 Computation . . . 92

7.3.3 Electrical Cabling . . . 92

7.4 Localized Manufacture . . . 94

A field as diverse as energy is potentially affected in many ways by nanotechnology, which has the opportunity to contribute in several ways to the diverse range of problems of the field. There are two important aspects: (1) how to supply small quantities of energy to nanoscale devices (e.g., implanted therapeutic devices, see Chapter 9)—we call this energy harvesting; and (2) how to address the current global undersupply of usable energy (and the trend is for the gap between demand and supply to increase)—we call this production and storage. Category (3) is more indirect, namely the multifarious contributions of nanotechnology to improving the energy efficiencies of human activities.

Two potentially—but still far from realization—important examples of category (3) have already been touched upon. One is the potential contribution of nanotechnology to metal extraction (Section 6.11). The other is the substitution of traditional building materials like concrete by ultrastrong (probably diamondoid) nanoplates (Section 6.5). Considering the very energy-intensive nature of concrete production, its elimination would greatly promote a favorable energy balance. A worthy goal of nanotechnology is, indeed, to enable all human requirements to be sourced renewably.

Applied Nanotechnology, Second Edition. http://dx.doi.org/10.1016/B978-1-4557-3189-3.00007-5

7.1 Energy harvesting

Solving this problem will make a negligible difference to the energy gap, but it will be enormously important in terms of convenience, on which the acceptance and takeup of large parts of nanotechnology ultimately depend. The devices currently being envisaged are typically energy transducers rather than heat engines in the conventional sense (e.g., a steam turbine). Most of the devices to be supplied are supposed to be worn by human beings. Therefore the energy sources are considered to be heat, mechanical movement, and chemical fuels. Heat harvesters would be based on pyroelectric materials positioned where significant heat gradients occur (e.g., the skin). Mechanical movement is of course ubiquitous but the efficiency of any kind of inertial device scales unfavorably with nanification [1], hence exploitation is likely to be located in zones where compression occurs (e.g., the soles of the feet). Energy-rich molecules such as glucose are sufficiently abundant in the blood for use as fuel, and fat reservoirs provide another potential source of chemical energy. Naturally, one will have to eat slightly more if energy is being tapped off for an artificial implant, but the difference is negligible. This would not apply if ambient radiant energy were harvested; the difficulty here is that the supply is typically irregular and hard to predict. The most exciting possibility is to exploit concentration gradients. This is of course how our cells harvest high-quality energy, especially using the remarkable—and very common—enzyme ATPase, which generates adenosine triphosphate (ATP) from transmembrane proton gradients [2].

Most energy harvesting research is at present taking place at the microscale, and is indeed an important part of the field of microsystems technologies (MST) or microelectromechanical systems (MEMS). The most actively pursued objectives are ways to power devices carried externally by human beings, such as cellphones and active clothing, for which the imperative to miniaturize down to the nanoscale is absent. Nevertheless, even a microscale energy harvester might usefully contain nanoscale components, hence nanotechnology will likely contribute to the field's development.

7.2 Production and storage

The Holy Grail of this field, which has inspired so much effort, is mimicry of natural photosynthesis.

Areas where nanotechnology might make a significant impact are photovoltaic cells and fuel cells. Natural photosynthesis (which is a combination of photovoltaic action with chemical storage) achieves the necessary photoinduced charge separation by extraordinarily sophisticated structure at the nanoscale. Regarding fuel cells, a major difficulty is the complex set of conflicting attributes that the materials constituting the cell, especially the important solid oxide type, must fulfil. Since nanocomposite materials are able to combine diverse attributes more effectively than conventional materials, there is some hope that more robust designs may emerge through a more systematic application of rational design exploiting nanomaterials.

7.2.1 Production

Solar energy. There is expected to be direct impact on photovoltaic cells converting radiant energy from the sun into electricity. The main primary obstacle to their widespread deployment is the high cost of conventional photovoltaic cells. Devices incorporating particles (e.g., Grätzel cells) offer potentially much lower fabrication costs, especially if inkjet printing technology can be adopted. The potential of incorporating further complexity through mimicry of natural photosynthesis, the original inspiration for the Grätzel cell, is not yet exhausted. The main challenge is to design and fabricate highly efficient and robust catalysts. This requires atomic control, however, and therefore must await the development of productive nanosystems (assembler-based fabrication). Robustness is a particularly difficult goal. Unlike natural systems, the components of which are constantly being renewed, the artificial system must remain functional for many years.

The main secondary obstacle to the widespread deployment of photovoltaic cells is that except for a few specialized applications (such as powering air-conditioners in Arabia) the electricity thus generated needs to be stored, hence the interest in simultaneous conversion and storage in chemical form, mimicking much more closely natural energy harvesting. This can also be encompassed within the concept of the Grätzel cell. Undoubtedly natural photosynthesis is only possible through an extremely exact arrangement of atoms within the photosystems working within plant cells, and the more precisely artificial light harvesters can be assembled, the more successful they are likely to be.

In view of the intensive research into all aspects of solar cells, the specific impacts of nanotechnology on progress might be quite difficult to discern. Reducing the thickness of some of the laminar elements in a photovoltaic cell will save on materials and may enhance efficiency. Reaction interfaces requiring catalysis will benefit from rationally designed catalysts constructed atom-by-atom, but there is no practical fabrication technology for that at present.

Presently photovoltaic electricity generation is rather expensive, and it has long managed to achieve a niche in the market only because of significant government subsidies. Two key parameters are cost and efficiency. The latter has been relentlessly increasing—albeit linearly rather than exponentially (see Figure 7.1)—but reducing the former has made giant strides more akin to exponential progress through the development of roll-to-roll printing of organic polymer solar cells. Presently these low-cost cells are less efficient and less robust but it seems to be only a matter of time before they become competitive with conventional means of generating electricity.

Government subsidies seem to have been largely responsible for significant solar panel manufacturing overcapacity in China. The firm Suntech, headquartered in Wuxi and listed in New York, was the world's largest solar panel maker in 2009. It was, however, alleged in Europe that solar panels from China were being sold for less

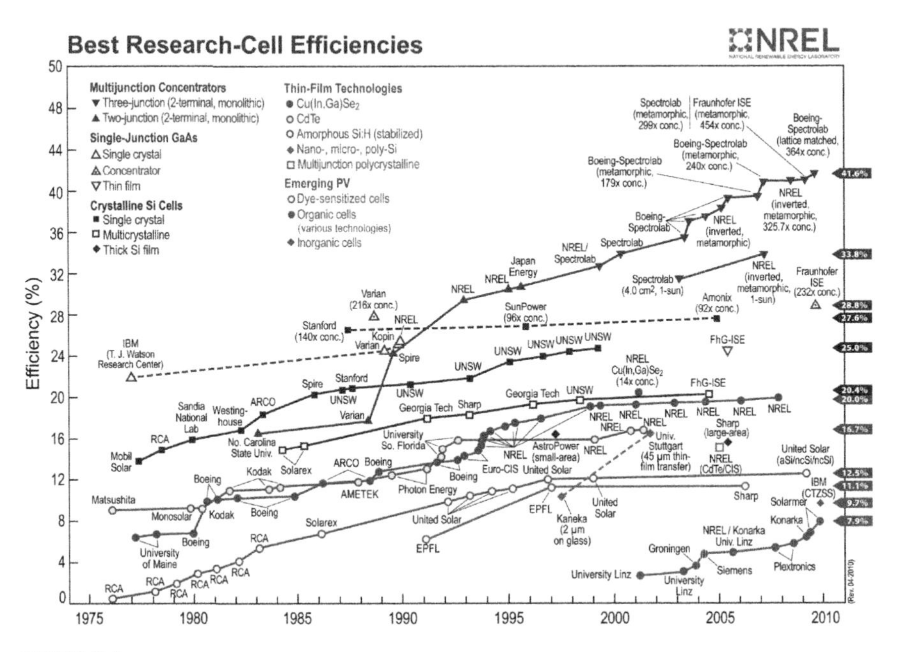

FIGURE 7.1

Evolution of research photovoltaic cell efficiencies for the different technologies currently under development. Data compiled by L. Kazmerski, National Renewable Energy Laboratory (NREL), Golden, CO. Reproduced from A.J. Parnell, Nanotechnology and the potential for a renewable solar future. *Nanotechnol. Perceptions* 7 (2011) 180–187, with permission from Collegium Basilea.

than their manufacturing costs. Be that as it may, the business model appears to be unsustainable: In March 2013, Suntech filed for bankruptcy protection.

Fuel cells. Although the scientific basis of this technology, whereby fuel is converted to electricity directly, was established over 160 years ago by Christian Schönbein, it has been very slow to become commercially established. As with photovoltaic cells, the main primary obstacle is the high cost of fabrication. Nanotechnology is expected to contribute through miniaturization of all components (especially reducing the thickness of the various laminar elements), simultaneously reducing inefficiencies and costs, and through realizing better catalysts for oxygen reduction and fuel oxidation. A priority is developing fuel cells able to use feedstocks other than ultrapure hydrogen. The catalytic problem is particularly acute here, since there seems to be an inverse relationship between the efficiency of a catalyst and its sensitivity towards inactivation by impurities in the fuel. Until we have routine atom-by-atom assembly, however, it seems that the impact of nanotechnology on catalyst fabrication will be minor since we shall not be able to much improve on existing technology.

Nanotechnology in the oil and gas industry. Although according to conventional wisdom the use of oil and gas as an energy source will diminish, its huge significance will remain for many more years, long enough for it to benefit from nanotechnology. Applications are extremely diverse. The extraordinary rheological properties of clay-based drilling muds have, in fact, been exploited for decades. There is much scope for even more advanced sensorial and responsive drilling fluids [3], for example basing them on polymers incorporating nano-objects. Many applications are generic, such as the use of nanocomposites as lightweight, rugged structural materials for drilling rigs, offshore platforms, pipelines, and so forth. Ultrahard nanocomposites can enhance the performance of drilling parts. Further downstream, nanocatalysts should enhance yields of petroleum refining operations, and precisely engineered nanomembranes may achieve convenient separation of desirable hydrocarbons from impurities. More indirectly, nano-engineered antifouling coatings for oil tankers have the potential to enable significant savings in the energy needed to propel them. Furthermore, the enormous expense of the current generation of giant tankers puts a premium on keeping them almost continuously in service, and their huge size makes servicing their hulls difficult, hence any contribution of nanotechnology to creating maintenance-free systems would be of great significance.

7.2.2 Storage

The primary means of storing energy is as fuel, but unless photoelectrochemical cells generating fuel from sunlight receive renewed impetus, renewable sources will mainly produce electricity and except for some special cases (such as photovoltaic-driven air conditioning installations in Arabia as already mentioned) at least some of the electricity will have to be stored to enable supply to match demand. High energy and power densities have become particularly relevant as a consequence of the search for replacements for the fossil fuel-powered internal combustion engine. The main contenders are supercapacitors and accumulators. The proliferation of portable electronic devices has also greatly increased demand for small and lightweight power sources for goods such as laptop computers and cellphones, especially in the absence of a viable personal energy harvesting technology.

Supercapacitors based on carbon nanotubes have attracted especial interest for rapidly responsive energy storage, but the impact of nanotechnology is likely to be small since using ordinary carbon black has already enabled over 90% of the theoretically maximum charge storage capacity using a randomly packed filler to be achieved, at much lower cost. In any case, the classic photoelectrochemical process generates hydrogen (from water), the storage of which is problematical. The same problem besets hydrogen-fed fuel cells—unless the storage problem is solved effectively and economically, the "hydrogen economy" will not be able to get established. Through the rational design and fabrication of storage matrix materials, nanotechnology should be able to contribute to effective storage, although whether this will tip the balance in favor of the hydrogen economy is still questionable. Conventional routes to materials synthesis have already achieved a high level of development without making use of nanotechnology, except in a secondary fashion by exploiting nanometrology tools during the discovery process.

Typically, energy storage devices such as batteries can be nanified by making internal layers thinner and more accurately, which not only reduces materials costs (provided they can be manufactured) but also improves their performance. The actual achievement, on an industrial scale, of nanoscale internal structure might well require a revolutionary change in manufacturing technology. Clearly, achieving it through assemblers would be revolutionary. Otherwise, the most revolutionary contribution seems to be through the use of nanoparticles rather than thin films. This enables well-established printing technologies to be used for fabricating devices.

7.2.2.1 *Electrical storage devices*

Electrical capacity depends on the distance separating the oppositely charged plates in a condenser, and interfacial area contributes to the performance of chemically fueled storage batteries. It is obvious that such systems can be structurally enhanced with respect to performance by nanification. A great deal of work is currently being undertaken in many laboratories worldwide (but especially in the USA, Japan, and some European countries) on these topics, and on the related one of hydrogen storage. The feasibility of implementing any technically appropriate solution will depend on issues such as compatibility and total lifecycle. Previously, it has been accepted that the relatively small achievable improvements in, say, supercapacitor performance through replacing carbon black, which is a very imperfect nanomaterial, by precisely formed carbon nanotubes were not worth the great increase in expense. However, significant improvements in carbon nanotube synthesis are constantly taking place and the feasibility of any proposed application will need to be assessed in terms of the very latest technology developments.

Table 7.1 compares the specific energies amd energy densities of some current and emerging technologies. A third parameter is specific power ("gravimetric power density"). Here, capacitors are as good as combustion fuels (around 1 MW/kg), which are much better than batteries and fuel cells (around 100 W/kg).

It will be noted that the Al-air battery (a rechargeable device developed by Europositron) has about the same practical output as the gasoline-fueled internal combustion engine. The best fuel cells are still operating far below their theoretical limits, but given that the field has already been researched for about 170 years it must be conceded that the barriers to further progress look to be considerable. Nanotechnology can in principle contribute in many ways. Lightweight refractory nanocomposites could increase the effective specific energy of gasoline, and nano-engineered catalysts would benefit fuel cells. Theoretically, nano-engineering will have the most direct impact on supercapacitors, whose stored energy depends directly on their internal surface area.

A bald list of specific energies might not be adequate to convey the potential of the technologies. Very often, it is through judicious combinations that really important benefits can be achieved. As an example, consider remote equipment such as miniature sensors continuously needing small amounts of electrical power. The combination of a small photovoltaic solar cell together with a supercapacitor capable of storing surplus

Table 7.1 Specific energies $\mathfrak{E}$ and energy densities $\mathfrak{D}$ of existing storage systems,[a] ranked by $\mathfrak{E}$.

Type	$\mathfrak{E}$(MJ kg^{-1})	$\mathfrak{D}$(GJ m^{-3})
Gasoline[b]	46	34
Aluminum[b]	31	84
Al-air battery	4.6	?
Thermite	4	18
Fuel cell[c]	1	wide range
Li-ion battery[c]	0.7	2
Na-NiCl battery[c]	0.43	?
Lead-acid battery[c]	0.14	0.36
Supercapacitor[c]	0.1	0.2

[a] *Specific energy is energy per unit mass—sometimes, confusingly, called energy density, which is energy per unit volume. The latter are less well defined than the former and are also liable to increase as new (nano)materials are developed.*
[b] *Not self-contained—requires external oxidizer and a combustion apparatus, the output from which might only be 10% of the given figure.*
[c] *Approximate, from laboratory experiments (there is a wide range depending on design and other considerations).*

energy during the day and releasing it at night allows the equipment to be completely autonomous regarding its energy requirements. This is the current state-of-the-art.

Supercapacitors are undergoing intense development. Rather than relying on random nanostructures (e.g., carbon black) *regular arrays* of nanofeatures may offer an increase of one to two orders of magnitude. This would put the supercapacitors ahead of the experimental Al-air battery and the practical output from gasoline. An intermediate, semistructured, development is to incorporate carbon nanotubes into cellulose-based paper [4]. Hitherto the main exploitation of nanotechnology for capacitors has been to increase the plate area. Nanoscale plate separation limits the voltage that can be stored on the plates because of arcing. There are now attempts to examine the theory more deeply, taking quantum effects explicitly into account, in order to understand how capacitor plates with separation distances in the nanoscale could be designed to avoid arcing at high voltages.

Obviously specific energies can only be compared if all other things are equal. These "other things" include all the infrastructure required to sustain the chosen system, including its cost, as well as charge/discharge characteristics, lifetime, etc. At least there seems to be no shortage of lithium as a raw material. Apart from efforts to increase power and energy densities, increasing the rapidity of charging and discharging would also be useful. Here too, regular nanostructures might be the key.

The novel nanomaterial graphene is currently being explored for a variety of energy storage applications. Clearly it is a candidate for capacitor plates, provided it can be stably configured. Carbon nanotubes are also finding novel applications, notably as high-capacity anodes in lithium ion batteries. A more exotic application is as a scaffold for fuel coatings which, when ignited, create a thermal (combustion) wave (thermowave) that entrains electrons as it moves along the carbon nanotube, generating current.

Apart from the above technologies, there is also design work being done on improving the nanoscale integration of the presently disparate components of electrical storage media.

7.2.2.2 Hydrogen storage

Hydrogen is attracting increasing attention as a fuel for automotive vehicles. Conventional hydrogen storage is either as a gas in a high pressure cylinder or as a liquid at low temperature [5]. For example, at 200 bar the density is approximately 10^{22} H-atoms/cm^3. In its liquid state (at 20 K) the density is fourfold that figure, and fivefold as a solid at 4 K. These figures exclude the sizes of the pressure vessel and the cryogenic equipment, the masses of which make it difficult to compute the gravimetric densities. Similar ambiguity attends the comparison of specific energies. Remarkably high densities can be achieved by storing the hydrogen in a metal as its hydride. For example, MgH_2 contains 6.5×10^{22} H-atoms/cm^3, although the fraction of the total mass that is hydrogen is only 0.076 (compared with about 0.04 for a high-pressure container). The main challenge in hydrogen storage is considered to be how to increase this mass fraction. High storage capacity is not, of course, the sole criterion; other operational parameters, such as charging and discharging rates, how much energy is required for charging (usually by high pressure) and discharging (usually by elevated temperature), and how many times it can be done before the structure deteriorates (the US Department of Energy specifies 1000 cycles), are also important. Many metals and alloys (typically of the general formulas AB_2 and AB_5) have been investigated. These have the advantages of operational pressures and temperatures within the practical ranges of 1–10 bar and 0–100 °C, unlike the more primitive substances such as MgH_2. $LaNi_5$ and its derivatives look very promising; hydrogen density at 2 bar, while not as good as that of MgH_2, equals that of gaseous molecular hydrogen at 1800 bar (this density is reduced by the packing fraction if the alloy is in powder form); adsorption is fast and reversible; cycling life is good. However, the achievable densities, even of MgH_2, are not as good as that of gasoline, which contains about 7×10^{22} H-atoms/cm^3. A variety of more exotic compounds capable of releasing hydrogen in the presence of a catalyst (e.g., formic acid), are being studied, but none are being seriously developed at present. An interesting new development is the "chemical hydrides," such as the Li–B–H and Li–N–H systems. Compared with the heavy metal alloys, these light elements yield much better hydrogen mass fractions (0.18 for $LiBH_4$). Most of this technology is chemistry. More relevant to nanotechnology is the possibility of using graphene or carbon nanotubes, or even relatively ill-characterized activated carbons, to store hydrogen. The achievable hydrogen storage densities are, however, rather on the low side ($\sim$0.04). Physisorption seems to be the mechanism of adsorption, a corollary of which is that rather low temperatures are required to get a reasonable amount of sorption.

In summary, hydrogen storage is a very active academic research field and the accumulated literature is already enormous. Little attention seems to be paid to the practicality of large-scale deployment of some of the more exotic materials. Reports summarizing new materials typically do not compare all relevant parameters, which include charging and discharging rates and energies apart from the overall storage capacity. The field is heavily driven by the automotive industry, the major players in which are carrying out intensive research.

As mentioned, the field is dominated by chemistry. The results obtained from exotic, highly structured materials (such as organosilicas) hint that rational nanomaterials design and assembly could yield storage materials with very attractive properties, although they could not at present be manufactured in the necessary large quantities. From that viewpoint, low-cost ways of making hierarchical porous materials based on well-established industrial processes (e.g., weaving electrospun polymers into sheets) seem to be rather attractive, with performance comparable to that of carbon nanotubes.

In general, nanification of the material, e.g., by having it in the form of nanoparticles, or by loading it into an aerogel of some inert material, will benefit the kinetics of charge and discharge, but typically will also result in less extreme environmental conditions (temperature and pressure) being required. The most exciting trends are those in which hydrogen responsivity is built into a material, such as the "pore-with-gate" metal-organic frameworks being developed at the University of Nottingham (UK), but this type of approach is still very much in its infancy.

7.3 Energy efficiency

This topic comprises a heterogeneous collection of technical impacts; most of them are incremental. Traditionally there are two opposing influences, the "utilities," which are mainly commercial operations trying to sell as much of their product (gas, electricity, etc.) as they sustainably can, and the consumer, who is trying to minimize costly energy consumption. Changes in generally held attitudes have resulted in most utilities helping consumers to minimize energy use. Since utilities are mostly local monopolies, they can offset reduced consumption by higher unit prices.

Nanostructured coatings with very low coefficients of friction and extremely good wear resistance will find application in all moving machinery, hence improving its efficiency and reducing the energy required to achieve a given result. Examples include electricity-generating wind turbines and electric motors. A similar example is the use of nanostructured surface coatings that can be painted on aircraft to reduce drag. The morphology of these coatings is typically biomimetic. Fuel savings of a few percent have been achieved in trials. The main difficulty is inadequate durability of the coatings. The materials themselves might not be particularly expensive, but the labor of applying them, especially if the aircraft has to be specially withdrawn from service, is costly.

Wind turbines should also benefit by having blades made from lightweight nanocomposites. The larger the blade, the greater the benefit.

7.3.1 Lighting and computation

Lighting appears here because it is an indispensable part of civilization, and is so widely used that improvements (e.g., more light output for the same input of electrical energy) have the potential to make a significant impact on energy consumption. For all applications where collateral heat production is not required, nanotechnology-enabled light-emitting diodes can replace incandescent filaments. Light-emitting semiconductor diodes can achieve a similar luminous output for much less power than incandescent filament lamps. The heat produced by the latter may be of value in a domestic context (e.g., contributing to space heating in winter) but is simply wasted in the case of outdoor street lighting (although the esthetic effect is agreeable) [6]. In fact, the actual operational efficiency of a device in a given functional context typically represents only a fraction of the overall effectiveness in achieving the intended function. For example, if the intended function of street lighting is to reduce road accidents, there is probably an ergonomic limit to the number of lamps per unit length of street above which the reduction becomes insignificant. Although data is hard to come by, it seems that few amenities have been properly analyzed in this manner. It may well be a 50% reduction in the number of lamps could be effected without diminution of the functional effect. Such a reduction would correspond to a really significant technological advance in the device itself.

Miniaturizing computer chips diminishes the heat dissipated per floating point operation (but, at present, not by as much as the increased number of floating point operations per unit area enabled by the miniaturization). Devices in which bits are represented as electron spins rather than as electron charges will dissipate practically no heat. If all digital information processors used single spin logic, which is most likely to be realized using nanotechnology [7], given the growing ubiquity of such processors, the contribution to energy economy would be very significant.

7.3.2 Computation

7.3.3 Electrical cabling

As long as electricity is produced at locations remote from its use, it must be transmitted along cables, in which process a considerable amount of energy is dissipated as heat. There is, therefore, much interest in lowering the resistance of electrical conductors.

Carbon nanotubes (CNTs) have, in certain forms, remarkable electrical attributes (Table 7.2). In the so-called "armchair" structure (in which the chiral vectors are equal) the band structure is metallic and the current capacity of the nanotubes is several orders of magnitude greater than the capability of copper (Table 7.2), in contrast to nanotubes having the so-called "zigzag" structure, which is semiconducting. The difficulty in creating a macroscopic electrical current-carrying cable, whose mass per unit length would be far smaller than that of an equivalent copper cable, is

Table 7.2 Some electrical properties of graphite and metallic carbon nanotubes,[a] compared with metals.

Property	unit	graphite[b]	CNT	composite	yarn	Cu	Al
Resistivity	Ω cm	0.1	5×10^{-6}[c]	> 100[g]	10^{-5}[d]	1.7×10^{-6}	2.7×10^{-6}
Curr. cap.	A cm^{-2}	20	2×10^{7}[d]	?	3×10^{4}[d]	600[h]	350[h]
Density	g cm^{-3}	2.1	2.5[f]	1	0.8	8.9	2.7

[a] *The given values are only approximate. Actual measured values still depend on many experimental details.*
[b] *Typically in the form of a carbon electrode.*
[c] *T.W. Ebbesen et al., Electrical conductivity of individual carbon nanotubes.* Nature *382 (1996) 54–55.*
[d] *T. Shimizu et al., Electrical conductivity measurements of a multiwalled carbon nanotube.* Surf. Interface Anal. *37 (2005) 204–207.*
[e] *Ref. [11]*
[f] *Multiwalled.*
[g] *The range of values is large. The polymer matrix without the nanotubes would typically have the resistivity of the order of* 10^{17}.
[h] *This is not very precisely defined, typically by when some destruction of the cable or its insulation takes place.*

essentially one of assembling and connecting up the nanotubes to make a macroscopic conductor. A single nanotube has a diameter of the order of 1 nm. Until relatively recently they could only be manipulated individually using the laborious procedures associated with scanning probe ultramicroscopies. An early approach was to randomly disperse them in a polymer matrix. Because of their extreme elongation, the percolation threshold is rather low (typically much less than 1 vol%) but these composite materials had disappointingly low conductivities (Table 7.2), presumably because each nanotube may only contact other nanotubes at two points and the contact resistance is high. Some improvement was noted when the nanotubes were clustered [8]. A significant breakthrough occurred with the demonstration of the ability to spin nanotubes into a yarn [9]. In such a material each nanotube may make hundreds or thousands of contacts with other nanotubes and the overall conductivity is correspondingly higher (Table 7.2). A single strand of yarn may have a diameter of some micrometers, and such strands may be braided or woven to create even larger structures.

Various improvements in the spinning process have been reported. For example, spinning from a dispersion of single-walled carbon nanotubes (SWNT) to which surfactant had been added resulted in a resistivity of 2.8×10^{-3} Ω cm [10]. Even more dramatic improvements were achieved by doping (double-walled) carbon nanotubes with iodine, enabling resistivities of as low as 10^{-5} Ω cm to be reached [11]. Another difficulty in realizing CNT-based cabling is the fact that known synthetic procedures for making the nanotubes produce a mixture of semiconducting and metallic ones, and of single-walled and multiwalled. One way of overcoming the problem is to refine the synthesis parameters. Methods of purifying the mixture are also being devised, including centrifugation and affinity separation using DNA.

7.4 Localized manufacture

Possibly the greatest ultimate contribution of nanotechnology, once the stage of the personal nanofactory has been reached, to energy conservation will be through the great diminution of the need to transport raw materials and finished products around the world. The amount of energy currently consumed by transport in one form or another is something between 30% and 70% of total energy consumption. A reduction by an order of magnitude is perhaps achievable.

References and notes

[1] Hierold C. From micro- to nanosystems: mechanical sensors go nano. J Micromech Microeng 2004;14:S1–S11.
[2] Yoshida M, et al. ATP synthase—a marvellous rotary engine of the cell. Nature Rev Mol Cell Biol 2001;2:669–77.
[3] Sensorial materials are materials in which the functions of sensing and structure are inseparably combined in a single material (Lawo M, et al. Simulation techniques for the description of smart structures and sensorial materials. J Biol Phys Chem 2009;9:143–8).
[4] Pushparaj VL, et al. Flexible energy storage devices based on nanocomposite paper. Proc Natl Acad Sci USA 2007;104:13574–7.
[5] The large electrolysis hydrogen production plant being built at Falkenhagen in Germany, in which hydrogen is simply a means of storing excess energy produced by renewable sources, especially wind turbines on very windy days, will slightly compress the gas and feed it into the natural gas distribution network.
[6] There has been much discussion about the psychological effects of different artificial light sources. The incandescent filament, which approximates to a black body and is therefore very natural, appears to correspond far better to a harmonious working environment than discharge lamps coated with fluorophores (i.e., what are commonly known as fluorescent tubes), the spectral emission profile of which is very different from that of a black body. Light-emitting diodes, which work according to the same principle (they are simply electroluminescent rather than photoluminescent), have a similar defect. Miniature nanoscale diodes could in theory be combined to create the same spectral output as the visible part of the black body radiation of an incandescent tungsten filament, but this does not seem to be a direction in which the technology is currently moving.
[7] Bandyopadhyay S, Single spin devices—perpetuating Moore's law. Nanotechnol Perceptions 2007;3:159–63.
[8] Aguilar JO, et al. Influence of carbon nanotube clustering on the electrical conductivity of polymer composite films. exPRESS. Polym Lett 2010;4:292–9.
[9] Zhang M, et al. Multifunctional carbon nanotube yarns by downsizing an ancient technology. Sci 2004;306:1358–61.

[10] Ma J, et al. Effects of surfactants on spinning carbon nanotube fibers by an electrophoretic method. Sci Technol Adv Mater 2010;11:065005.

[11] Zhao Y, et al. Iodine doped carbon nanotube cables exceeding specific electrical conductivity of metals. Scientific Reports 1:83 (doi: 10.1038/srep00083).

CHAPTER

Information Technologies

8

CHAPTER OUTLINE HEAD

8.1 Silicon Microelectronics 98
8.2 Heat Management 98
8.3 Data Storage Technologies 99
8.4 Display Technologies 100
8.5 Molecule or Particle Sensing Technologies 100

It has already been pointed out that for information processing (including data storage) applications, nanotechnology offers many advantages because the intrinsic lower limit of the representation of one bit of information is around the atomic (nano) scale. The process of nanification of information processing technology is well represented by Moore's law—which in its original form makes the empirical assertion that the number of components (i.e., resistors, capacitors, and resistors) per chip doubles each year (a manifestation of vastification) [1]. When Moore revisited this prediction 10 years later [2], he somewhat refined his statement (since about 1972 up to the present, the number of transistors per chip has roughly doubled every 2 years), pointing out that the result was the consequence of three technological developments: increasing area per chip, decreasing feature size (defined as the average of linewidth and spacewidth), and improved design of both individual devices and the circuit. Only the second of these three is a nanification process.

The direct economic consequence of these technological developments is a roughly constant cost per area of processed silicon, while the processing power delivered by the chip becomes ever greater. Furthermore, nanification makes the transistors not only smaller, but also lighter in weight, faster (because the electrons have less distance to travel), less power-hungry and more reliable. These are all strong selling points. Therefore, although technology push is undoubtedly important in maintaining Moore's law, the ultimate driver is economic.

As a result of these developments the microprocessor, which nowadays contains nanoscale components [3], has become ubiquitous throughout the world. For example, even a small company employing fewer than 50 people probably uses a computer to

Applied Nanotechnology, Second Edition. http://dx.doi.org/10.1016/B978-1-4557-3189-3.00008-7

administer salaries, etc. (even though it would almost certainly be cheaper and more effective to do it manually).

8.1 Silicon microelectronics

The starting point of chip production is the so-called wafer, the circular disk cut from a single crystal of silicon with a diameter of at least 300 mm and a thickness typically between 500 and 800 μm. Using lithography and etching technology the structures of integrated circuits are built up layer-by-layer on the surface of a chip [4]. Transistor construction has been based on complementary metal oxide–semiconductor (CMOS) technology for decades. The size of the smallest structures has been steadily diminishing—45 nm in 2009, 32 nm in 2012, and 22 nm in 2015 according to the transistor "roadmap"—this last value is close to the operational limit for metal oxide semiconductor field-effect transistor technology. These developments represent tremendous technological challenges, not only in the fabrication process itself, but also in testing the finished circuits and in heat management—a modern high-performance chip may well dissipate heat at a density of 100 W/cm^2, greater than that of a domestic cooking plate.

Silicon itself is still foreseen as the primary semiconducting material (although germanium, gallium arsenide, etc. continue to be investigated), but in order to fabricate ever-smaller structures, new photoresists will have to be developed. Furthermore, the silicon oxide thin film, which insulates the gate from the channel in the field-effect transistor, becomes less and less effective as it becomes thinner and thinner (of the order of 1 nm). Other metal oxides (e.g., hafnium oxide) are being investigated as alternative candidates.

Some design issues arising from this relentless miniaturization are discussed in Chapter 15.

8.2 Heat management

Heat management has grown in importance with the nanification of electronics, because the concomitant increase in the number of components per unit chip area implies a parallel increase in the heat generated (by the passage of electron currents).

Since the surface of the object emitting heat inevitably has a certain roughness, the contact with any juxtaposed heat sink will be imperfect. As in tribology, only the contact asperity is relevant, and its area may be quite small. The purpose of *thermal interface materials* (TIM) is to fill the gaps. These materials are typically pastes of highly conducting nano-objects (carbon nanotubes are especially useful—cf. Table 6.1) dispersed in silicone grease. A 10% volume fraction of the filler can increase thermal conductivity of the material from about 0.2 $W\,m^{-1}\,K^{-1}$ to almost 2 $W\,m^{-1}\,K^{-1}$. The use of TIMs can typically diminish the interfacial thermal resistance by an order of magnitude (around 5 mm^2 K/W is achievable). TIMs are essentially a kind of nanofluid that is preferably solid when cold.

Carbon nanotube arrays also exploit the extremely high thermal conductivity of carbon nanotubes. They can be synthesized directly on the silicon or silica substrate and perpendicular to it, forming a very efficient bridge between the electronic component and the heat sink [5].

8.3 Data Storage Technologies

Electrons have spin as well as charge. This is of course the origin of ferromagnetism and, hence, magnetic memories, but their miniaturization has been limited not by the ultimate size of a ferromagnetic domain but by the sensitivity of magnetic sensors. In other words, the main limitation has not been the ability to make very small storage cells, but the ability to detect very small magnetic fields.

The influence of spin on electron conductivity was invoked by Nevill Mott in 1936, but remained practically uninvestigated and unexploited until the discovery of giant magnetoresistance (GMR) in 1988. The main current application of spintronics (loosely defined as the technology of devices in which electron spin plays a rôle) is the development of ultrasensitive magnetic sensors for reading magnetic memories. Spin transistors, in which the barrier height is determined by controlling the nature of the electron spins moving across it, and devices in which logical states are represented by spin, belong to the future (Chapter 16).

Giant magnetoresistance (GMR) is observed in thin (a few nanometers) alternating layers (superlattices) of ferromagnetic and nonmagnetic metals (e.g., iron and chromium) [6]. Depending on the width of the nonmagnetic spacer layer, there can be a ferromagnetic or antiferromagnetic interaction between the magnetic layers, and the antiferromagnetic state of the magnetic layers can be transformed into the ferromagnetic state by an external magnetic field. The spin-dependent scattering of the conduction electrons in the nonmagnetic layer is minimal, causing a small resistance of the material, when the magnetic moments of the neighboring layers are aligned in parallel, whereas for the antiparallel alignment the resistance is high. The technology is nowadays used for the read–write heads in computer hard drives. The discovery of GMR depended on the development of methods for making high-quality ultrathin films (such as molecular beam epitaxy).

A second type of magnetic sensor is based on the magnetic tunnel junction (MTJ), in which a very thin dielectric layer separates ferromagnetic (electrode) layers, and electrons tunnel through the nonconducting barrier under the influence of an applied voltage. The tunnel conductivity depends on the relative orientation of the electrode magnetizations and the tunnel magnetoresistance (TMR): it is low for parallel alignment of electrode magnetization and high in the opposite case. The magnetic field sensitivity is even greater than for GMR. MTJ devices also have high impedance, enabling large signal outputs. In contrast with GMR devices, the electrodes are magnetically independent and can have different critical fields for changing the magnetic moment orientation. The first laboratory samples of MTJ structures (NiFe–Al_2O_3–Co) were demonstrated in 1995.

8.4 Display technologies

The results of a computation must, usually, ultimately be displayed to the human user. Traditional cathode ray tubes have been largely displaced by the much more compact liquid crystal displays (despite their disadvantages of slow refresh rate, restricted viewing angle, and the need for back lighting). The main current rival of liquid crystal displays are organic light-emitting diodes (OLEDs). They are constituted from an emissive (electroluminescent), conducting organic polymer layer placed between an anode and a cathode (see also Chapter 13).

Any light-emitting diode requires one of the two electrodes to be transparent. Traditionally, indium-doped tin oxide (ITO) has been used, but the world supply of indium is severely limited, and at current rates of consumption may be completely exhausted within 2 or 3 years. Meanwhile, relentless onward miniaturization and integration make it more and more difficult to effectively recover indium from discarded components. Hence there is great interest in transparent polymers doped with a small volume percent of carbon nanotubes to make them electrically conducting (see Section 6.2).

8.5 Molecule or particle sensing technologies

Information technology has traditionally focused on arithmetical operations, but information transduction belongs equally well to the field. Information represented as the irradiance of a certain wavelength of light, or the bulk concentration of a certain chemical, can be converted (transduced) into an electrical signal. From careful consideration of the construction of sensors consisting of arrays of discrete sensing elements, it can be clearly deduced that atomically precise engineering will enable particle detection efficiency to approach its theoretical limit [7]. Since a major application of such sensors is to clinical testing, they are considered again in the next chapter.

References and notes

[1] Moore GE. Cramming more components onto integrated circuits. Electronics 1965;38: 114–17.

[2] Moore GE. Progress in digital integrated electronics. In: International electron devices meeting (IEDM) technical digest. Washington, DC; 1975. p. 11–13.

[3] Nowadays, the sizes of apparatus such as a cellphone or a laptop computer are limited by peripherals such as screen, keyboard, and power supply, not by the size of the information processing unit.

[4] Mamalis AG et al. Micro and nanoprocessing techniques and applications. Nanotechnol Perceptions 2005;1:63–73.

[5] Son Y et al. Thermal resistance of the native interface between vertically aligned multiwall carbon nanotube arrays and their SiO_2/Si substrate. J Appl Phys 2008;103:024911.
[6] Baibach MN et al. Giant magnetoresistance of (001)Fe/(001)Cr magnetic superlattices. Phys Rev Lett 1988;61:2472–5.
[7] Manghani S, Ramsden JJ. The efficiency of chemical detectors. J Biol Phys Chem 2003;3:11–17.

CHAPTER

Health

9

CHAPTER OUTLINE HEAD

9.1 Current Activity 104
9.1.1 Nano-Objects 104
9.1.2 Other Nanomaterials 106
9.1.3 Information Technology 106
9.2 Longer-Term Trends 107
9.3 Implanted Devices 107
9.4 Paramedicine 109

Nanomedicine is defined as the application of nanotechnology to human health. The dictionary definition of medicine is "the science and art concerned with the cure, alleviation and prevention of disease, and with the restoration and preservation of health." As one of the oldest of human activities accessory to survival, it has made enormous strides during the millennia of human civilization. Its foremost concern is well captured by Hippocrates' dictum, "Primum non nocere." During the past few hundred years, and especially during the past few decades, medicine has been characterized by enormous technization, with a concomitantly enormous expansion of its possibilities for diagnosing and curing diseases. The application of nanotechnology, the latest scientific–technical revolution, is a natural continuation of this trend.

Medicine is, of course, closely allied to biology and, as already mentioned, molecular biology can well be considered an example of conceptual nanotechnology—scrutinizing a system with nanoscale resolution. Furthermore, much of the actual work of the molecular biologist increasingly involves nanometrology, such as the use of scanning probe microscopies.

Much closer to nanotechnology is the mimicry, by artificial means, of natural materials, devices, and systems structured at the nanoscale. Ever since Drexler presented biology as a "living proof of principle" of nanotechnology [1], there has been a close relationship between biology and nanotechnology.

It is customary nowadays to take a global view of things, and in assessing the likely impact of nanotechnology on medicine this is very necessary. Nanotechnology

Applied Nanotechnology, Second Edition. http://dx.doi.org/10.1016/B978-1-4557-3189-3.00009-9

is often viewed to be the key to far-reaching social changes (this theme will be taken up again in Chapter 18), and once we admit this link then we really have to consider the gamut of major current challenges to human civilization, such as demographic trends (overpopulation, aging), climate change, pollution, exhaustion of natural resources (including fuels), and so forth. Nanotechnology is likely to influence many of these, and all of them have implications for human health. Turning again to the dictionary, the definition of medicine continues to include "the art of restoring and preserving health by means of remedial substances and the regulation of diet, habits, etc." It would be woefully inadequate if the impact of nanotechnology on medicine were restricted to consideration of the development of more sophisticated ways of packaging and delivering drugs (important as that is).

Nanomedicine appears to have begun its era of exponential growth. Over the decade 1990–2001 the increase in the number of scientific publications in the field looks roughly linear, as does the increase in the annual number of patents filed. After 2000 the growth curves significantly up and away from the linear trend [2]. Nanotechnology-based drug delivery started its exponential phase around the year 2006, before which the market was negligible. By 2010 it had already reached a size of $1000 million.

The oldest well-documented example of the use of nanoparticles in medicine is perhaps Paracelsus' deliberate synthesis of gold nanoparticles (called "potable gold") as a pharmaceutical preparation [3]. This is an excellent example of reformulating a drug (gold) in order to improve its bioavailability.

We may conveniently classify nanomedicine products temporally: the first group is more "micro" than "nano" and is already well established, including technologies such as laparoscopic diagnosis and surgery with the help of microscale instruments, and the *in vitro* tests carried out on various kinds of "lab-on-a-chip," which only require minute quantities of sample (e.g., blood) compared with conventional analytical technology, and are cheap and portable enough for such devices to be found not in central analytical laboratories but in the general practitioner's surgery and even nowadays in the private home. These products will not be further considered in this book.

The second group (Section 9.1) has been well researched and is currently under intensive development, with commercial products just starting to emerge, although they cannot yet be said to be well established. This group can in turn be divided into nano-objects, other nanomaterials, and nano-enabled information processing.

The third group comprises longer-term developments envisaged by the pharmaceutical industry (Section 9.2).

9.1 Current activity

9.1.1 Nano-objects

All the medical applications of nanoparticles and other nano-objects have one feature in common: the objects must be functionalized to confer on them a specific affinity for a biological target, which may be a certain type of cell, in which case the specific target might be a particular protein expressed on the surface of the cell. In use, the nano-objects are introduced systemically into the body (usually the bloodstream), typically in very large numbers, and thanks to their specific affinity they will end up being concentrated at the target site(s).

For diagnostic imaging applications, the particle should have good contrast compared with the background tissue for whatever microscopy is used subsequently to examine the tissue. Inorganic fluorophores (quantum dots) are useful, not least because the wavelength of their fluorescent emission can be easily tuned simply by slightly changing the size of the dot, enabling multiple diagnoses to be carried out simultaneously. Gold and similar metals are useful for electron microscopy, and so forth.

If the particles are superparamagnetic, they can be heated by the external application of an oscillating magnetic field, killing cells in their vicinity. The treatment has been called magnetic fluid hyperthermia. The main envisaged application is for killing tumors, especially small metastases that are extremely difficult to detect, let alone eliminate, by conventional means. If a large tumor of known location is to be eliminated, it would be possible to steer such particles using external magnetic fields, firstly to the target site, and then to the gastrointestinal tract for excretion after they had done their work.

A vast and diverse domain of application is to use nano-objects as drug delivery vehicles. Many therapeutically effective substances are useless as drugs if they are simply administered in the conventional manner, because they fail to penetrate the numerous biophysical obstacles in the body (e.g., cell or intracell membranes), or are destroyed by the immune system, and so forth. Nano-objects can be made undetectable to the immune system (the best strategy appears to be to make them superhydrophilic) and these stealthy or simply "stealth" objects can then be used to carry drugs without the danger of opsonization and elimination. The nano-object requires some kind of internal reservoir into which the drug can be loaded and, preferably, some kind of activation mechanism that will enhance release of the drug when the particle reaches its target. It is a great challenge to combine all this functionality into a single nano-object. A simple example of a "smart" (probably it is more accurate to describe it as merely "responsive") drug delivery particle is a hollow nanoshell of calcium carbonate destined for the stomach: the strongly acidic environment there will dissolve the mineral shell away, releasing the contents. In that case there is no problem about the subsequent fate of the nanoparticles, but many materials are not decomposed in such a fashion and their subsequent fate may be problematical. Ideally they should be excreted via the gastrointestinal tract after they have delivered their load but, depending on their size, they may penetrate inside cells (e.g., those of the liver) and lodge there more or less permanently, with presently unknown physiological consequences.

It would seem to be a relatively small step from an advanced nano-object capable of reaching its target, responsively releasing its contents, and then heading for the gastrointestinal tract to a nanoscale robot or "nanobot" capable of carrying out even more sophisticated operations. Microscopic or nanoscopic robots can also be considered to be an extension of existing ingestible devices that slowly move through the gastrointestinal tract and gather information (mainly images). The nanobot would need an internal energy source (although it may be able to harvest energy from its immediate environment), the ability to receive and transmit information (possibly to congeners), and some information processing capability. In effect it would start to approach the functionality of a living cell [4]. Minimal capabilities required of future devices are: (chemical) sensing; communication (receiving information from, and transmitting information to, outside the body, and communication with other nanobots); locomotion—operating at very low Reynolds numbers, estimated at about 1/1000 (i.e., viscosity dominates inertia); computation (e.g., recognizing a biomarker would typically involve comparing sensor output to some preset threshold value; due to the tiny volumes available, highly miniaturized molecular electronics would be very attractive for constructing on-board logic circuits); and of course power—it is estimated that picowatts would be necessary for propelling a nanobot at a speed of around 1 mm/s. It is very likely that to be effective, these nanobots would have to operate in swarms, putting an added premium on their ability to communicate.

9.1.2 Other nanomaterials

The three main categories are: tissue scaffolds for enabling artificial tissues to be grown *in vitro* for subsequent incorporation into a patient's body; structural implants (e.g., nanocomposites for repairing teeth [5,6]) and implanted devices (Section 9.3). Concerning tissue scaffolds, it is now known that the extracellular matrix (ECM), which acts as a scaffold in the body on which cells grow, has a complex structure made up of several different kinds of large protein molecules (e.g., laminin, tenascin). The responses of cells in contact with the ECM have revealed dramatic changes correlated with subtle differences in the molecules. The main research question at present is to determine what features of artificially nanostructured substrata are required to induce a cell to differentiate in a certain way. An enormous quantity of results has already been accumulated, but overall it seems not to have been sufficiently critically reviewed, and therefore it is difficult at present to discern guiding principles, other than rather trivial ones such as cells aligning themselves along grooves.

9.1.3 Information technology

The computer has become pervasive in the pharmaceutical industry: in research (especially computer modeling of the docking of drugs to putative biological targets); in development (driving laboratory automation, especially for high throughput screening); and in production (reactor control and tagging batches of chemicals). Personal computers are also ubiquitous in the surgeries of general practitioners, where they might be used to rapidly access information (from the internet) and for simple word-processing tasks such as writing out a prescription. Hospitals have the more ambitious aim of storing patient data in electronic form, which has, however, seen some spectacular failures in recent years. Possibly the largest-scale failure ever in the history of software engineering was that of the collapse of the "Connecting for Health" project that was intended to equip the National Health Service in England with a common computing infrastructure [7]. The use of the computer for carrying out computations (pattern recognition) for automated diagnosis is a further developing area.

Apart from these applications, real computations are also, of course, essential for constructing the images that emerge from X-ray tomography and similar techniques, but the computations are internal and the physician will generally confine his or her attention to the image that is produced.

9.2 Longer-term trends

The three main long-term developments envisaged by leading pharmaceutical companies are: sensorization [8], automated diagnosis, and customized pharmaceuticals.

Sensorization belongs predominantly to direct nanotechnology. With their ever-diminishing footprint, nanoscale sensors are not only able to penetrate ainside the body via minimally invasive procedures such as endoscopy, but are moving toward the ability to be permanently implanted. The latter are potentially capable of yielding continuous outputs of physiologically relevant physicochemical parameters such as temperature and the concentrations of selected biomarkers. Compared with the conventional practice of carrying out a blood test once or twice a year, the result of which is essentially a random snapshot of the instantaneous physiological state of the patient, having continuous data is an enormous advance. The downside of these developments is the enormous quantity of data that needs to be handled, but here indirect nanotechnology comes to the rescue, with ever-increasing information processing power becoming available.

One of the greatest current challenges is the automatic diagnosis of disease. If indeed "about 85% of [medical examination] questions require only recall of isolated bits of factual information" [9], this looks to be achievable even by currently available computing systems.

The third development, customized pharmaceuticals, is driven by the growing realization that human genetic polymorphism, especially of the cytochrome P450 enzymes, brings great variety to drug responses among individual patients [10]. Affordable customized pharmaceuticals are supposed to be enabled by miniaturized (microfluidic) mixers and reactors, but this is micro rather than nano (although it may be nano-enabled [11]) and, hence, outside the scope of this book.

9.3 Implanted devices

Prostheses and biomedical devices must be biocompatible [12]. This attribute can take either of two forms: (i) implants fulfilling a structural rôle (such as bone replacements) must become assimilated with the host; failure of assimilation typically means that the implant becomes coated with a layer of fibrous material within which it can move, causing irritation and weakening the structural rôle; (ii) for implants in the bloodstream (simple passive devices such as stents, and implanted sensors in the future) the opposite property is required: blood proteins must not adsorb on them.

Adsorption has two deleterious effects: layers of protein buildup and may clog the blood vessel; or the proteins that adsorb may become denatured, hence foreign to the host organism and triggering inflammatory immune responses.

In order to promote assimilation, a favorable nanotexture of the surface seems to be necessary, to which the cell responds by excreting extracellular matrix molecules, "humanizing" the implant surface. Years of empirical studies have enabled this to be achieved in some cases (e.g., the surfaces of cell culture flasks). Intensive research is meanwhile under way to provide the basis for a more rational design of the surfaces with a pattern specified at the atomic level; it is still not known whether the pattern only needs to fulfil certain statistical features.

In order to prevent adsorption, its free energy (ΔG) is analyzed according to [13]:

$$\Delta G_{123} = \Delta G_{22} + \Delta G_{13} - \Delta G_{12} - \Delta G_{23}, \tag{9.1}$$

where subscripts 1, 2, and 3 denote the implant (surface), the biofluid (blood), and the protein, respectively. ΔG_{22} is thus the cohesive energy of water, which is so large that this term alone will ensure that adsorption occurs unless it is countered by strong hydration. The biomedical engineer cannot influence ΔG_{23} and must therefore design ΔG_{12} appropriately. Coating material 1 with an extremely hydrophilic material such as poly(ethylene oxide) (PEO) is one way of achieving this.

For medical devices that are not implanted, such as scalpels or needles, attention is paid to finishing them in such a way that they cut the skin very cleanly, minimizing pain, and have low coefficients of friction to allow penetration with minimal force [14]. This may be achieved by ultraprecision machining, finishing the surfaces to nanometer-scale roughness.

Long-term implants must be designed in such a way as not to host adventitious infection by bacteria. Once they colonize an implant, their phenotype usually changes and they may be resistant to the attentions of the body's immune system (thus causing persistent inflammation without being destroyed) and to antibiotics.

Implants with rubbing surfaces, such as joint replacements, typically generate particles as a result of wear. Traditional tribopairs such as metal–polyethylene generate relatively large microparticles (causing inflammation); novel nanomaterials with otherwise improved properties may generate nanoparticles, with unknown consequences.

Nanotechnology is thought to be attractive for substance-sensing devices because, it is felt, if the device is comparable in size to the object being sensed it will be extremely sensitive to the presence of that object. For example, the conductivity of a fragment of graphene may dramatically change if a single oligopeptide is adsorbed onto it. Whether such physical effects can be transformed into reliable sensing devices capable of providing the ambient concentration of the substance of interest does not seem to have been given much attention. The photon is probably the smallest entity, the concentration of which is routinely measured, and for well understood reasons the more sensitive the measurement demand, the bigger the detector (as astronomers well know). For a substance sensor to be effective, it is the *molecular detection efficiency* that needs to be optimized [15].

9.4 Paramedicine

The use of toxic materials for cosmetic purposes (e.g., applying them to the skin of the face) has a long history—antimony salts were popular among the Romans, for example. The data given in Table 5.3 testify to the popularity of nanotechnology in this area. Advances in our knowledge of toxicity have since then ushered in far more benign materials, although the recent use of extremely fine particles (for example, zinc oxide nanoparticles in sunscreens) has raised new concerns about the possibility of their penetration through the outer layers of the skin, or penetration into cell interiors, with unknown effects. Many modern cosmetic products are amazingly sophisticated in their nanostructure. An important goal is to devise delivery structures for poorly water-soluble ingredients such as vitamin A and related compounds, vitamin E, lycopene, and so forth. The liposome (a lipid bilayer enclosing an aqueous core, i.e., a vesicle) is one of the most important structures; the first liposome-based cosmetic product was launched by Dior in 1986. The existence of variants such as "transferosomes" (liposomes with enhanced elasticity), "niosomes" (using nonionic surfactants instead of the lipid bilayer), "nanostructured lipid carriers," "lipid nanoparticles," and "cubosomes" (fragments of the bicontinuous cubic phase of certain lipids) point to the intense development activity in the field, mostly carried out in-house by the firms.

References and notes

[1] Drexler KE. Molecular engineering: an approach to the development of general capabilities for molecular manipulation. Proc Natl Acad Sci USA 1981;78:5275–8.

[2] Maitra A. Nanotechnology and nanobiotechnology. Nanotechnol Perceptions 2010;6:197–204.

[3] Zsigmondy R, Thiessen PA. Das kolloide Gold. Leipzig: Akademische Verlagsgesellschaft; 1925.

[4] Hogg T. Evaluating microscopic robots for medical diagnosis and treatment. Nanotechnol Perceptions 2007;3:63–73.

[5] Roveri N, et al. Recent advancements in preventing teeth health hazard: the daily use of hydroxyapatite instead of fluoride. Recent Patents Biomed Engng 2009;2:197–215.

[6] Moszner N, Klapdohr S. Nanotechnology for dental composites. Intl J Nanotechnol 2004;1:130–56.

[7] Sampson G. Whistleblowing for health. J Biol Phys Chem 2012;12:37–43.

[8] Defined as embedding large numbers of sensors in a structure, in this case the human body.

[9] According to a University of Illinois study by G. Miller and C. McGuire (quoted by Fabb WE. Conceptual leaps in family medicine: are there more to come? Asia Pacific Family Med 2002;1:67–73).

[10] Kalow W et al. Multigenic traits and risk assessment in pharmacology: a population approach. Pharmacogenomics J 2001;1:234–6.

[11] Iles A. Microsystems for the enablement of nanotechnologies. Nanotechnol Perceptions 2009;5:121–33.

[12] Ramsden JJ. Biomedical Surfaces. Norwood, MA: Artech House; 2008.
[13] Cacace MG, et al. The Hofmeister series: salt and solvent effects on interfacial phenomena. Q Rev Biophys 1997;30:241–78.
[14] Ramsden JJ, et al. The design and manufacture of biomedical surfaces. Annals CIRP 2007;56(2):687–711.
[15] Manghani S, Ramsden JJ. The efficiency of chemical detectors. J Biol Phys Chem 2003;3:11–17.

Further reading

[1] Freitas Jr RA. What is nanomedicine? Nanomed Nanotechnol Biol Med 2005;1:2–9.
[2] Kubik T et al. Nanotechnology on duty in medical applications. Current Pharmaceutical Biotechnol 2005;6:17–33.

PART

ORGANIZING NANOTECHNOLOGY BUSINESS

III

CHAPTER

The Business Environment 10

CHAPTER OUTLINE HEAD

10.1 The Universality of Nanotechnology 113
10.2 The Radical Nature of Nanotechnology 116
10.3 Intellectual Needs 117
10.4 Company–University Collaboration 119
10.5 Clusters 120
10.6 Assessing Demand for Nanotechnology 120
10.6.1 Modeling 121
10.6.2 Judging Innovation Value 121
10.6.3 Anticipating Benefit 122
10.7 Technical and Commercial Readiness (Availability) Levels 123
10.8 Predicting Development Timescales 125
10.9 Nanometrology 127
10.10 Standardization of Nanotechnology 129
10.11 Patents 130

Factors considered to contribute to the success of the nanotechnology industry are considered in this chapter, apart from the fiscal environment (including geography) and regulation, which are placed in separate chapters following this one. Chapter 13 contains some company case studies.

10.1 The universality of nanotechnology

Reference is often made to the diversity of nanotechnology. Indeed, some writers insist on referring to it in the plural as nanotechnologies (perhaps an unnecessarily refined nuance). Inevitably, a technology concerned with building matter up atom-by-atom is a universal technology with enormous breadth [1], which can be applied to virtually any manufactured artifact. Nanostructured materials are incorporated into nanoscale devices, which in turn are incorporated into many products, as documented in Part II. An artifact is considered to be part of nanotechnology if it contains nanostructured

Applied Nanotechnology, Second Edition. http://dx.doi.org/10.1016/B978-1-4557-3189-3.00010-5

materials or nanoscale devices even if the artifact itself is of microscopic size; this is the domain of indirect (or enabling) nanotechnology (Section 1.4). The fact that the feature sizes of components on semiconductor microprocessor chips are now smaller than 100 nm, and hence within the nanoscale, means that practically the entire realm of information technology is now becoming part of nanotechnology. Nanotechnology is, therefore, becoming pervasive [2]. The best extant example of such a universal technology is probably information technology, which is used in countless products.

Any universal technology—and especially one that deals with individual atoms directly—is almost inevitably going to be highly upstream in the supply chain. This is certainly the case with nanotechnology at present. Only in the case of developments whose details are still too nebulous to allow one to be anything but vague regarding the timescale of their realization, such as quantum computers and general-purpose atom-by-atom assemblers, would we have pervasive direct nanotechnology.

Universal technologies form the basis of new value creation for a broad range of industries; that is, they have "breadth." Such technologies have some special difficulties associated with their commercialization because of their upstream position, far from the ultimate application (see Figure 10.1).

The most important difficulty is that the original equipment manufacturer (OEM) needs to be persuaded of the advantage of incorporating the nanoscale component or nanomaterial into the equipment. The most convincing way of doing this is to construct a prototype. But if the technology is several steps upstream from the equipment, constructing such a prototype is likely to be hugely expensive (presumably it will anyway be outside the domain of expertise of the nanotechnology supplier, so will

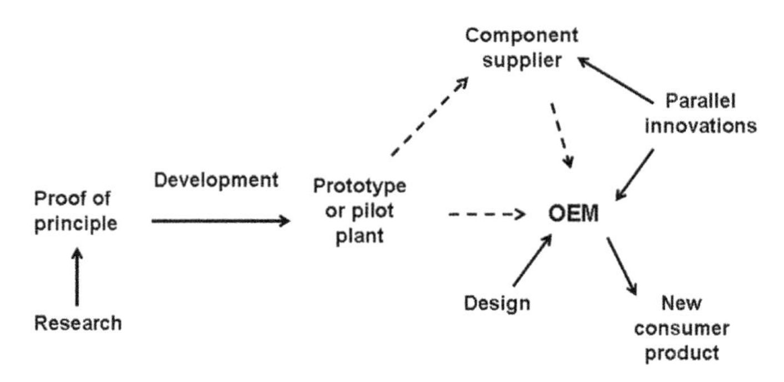

FIGURE 10.1

Diagram of immediate effects showing the supply chain from research to consumer product. The dashed lines indicate optional pathways: the route to the original equipment manufacturer (OEM) is very likely to run via one or more component suppliers. Parallel innovations may be required for realization of the equipment. These include legally binding regulatory requirements (particularly important in some fields—e.g., gas sensors). Note that most of the elements of M. Porter's value chain are included in the last arrow from OEM to consumer product.

have to be outsourced). The difficulty is compounded by the fact that many OEMs, especially in the important automotive branch, as well as "Tier 1" suppliers, rarely pay for prototype development. The difficulty is even greater if a decision to proceed is taken at the ultimate downstream position, that of the consumer product itself. The nanotechnology supplier, which as a start-up company is typically in possession only of proof of principle, often obtained from the university laboratory whence the company sprang, is likely to be required to make its most expensive investments (e.g., for a prototype device or an operational pilot plant) before it has had any customer feedback, which is so important for determining the cost-effectiveness.

This distance between the technology and its ultimate application will continue to make life difficult for the technologist even if the product containing his technology is introduced commercially, because the point at which the most valuable feedback is available—from the consumer—is so far away. There is perhaps an analogy with Darwin's theory of evolution here, in its modern interpretation that incorporates knowledge of the molecular nature of the gene—variety is introduced at the level of the genome (e.g., via mutations), but selection operates a long way downstream, at the level of the organism. The disparity between loci is especially acute when the exigencies of survival include responses to potentially fatal sudden threats.

The further upstream one is, the more difficult it is to "capture value" (i.e., generate profit) from one's technology [3]. Hence cash tends to be limited, which in turn limits the possibilities for financing demonstration prototypes.

The difficulty of the position of the upstream technologist is probably as low as it can be if the product or process is one of substitution. As will be seen later (Chapter 13) this is likely to be a successful path for a small nanotechnology company to follow. In this case, demonstration of the benefits of the nanotechnology is likely to be relatively straightforward, and might even be undertaken by the downstream client.

On the other hand, the nanotechnology revolution is unlikely to be realized merely by substitutions. Much contemporary nanotechnology is concerned with a greater innovative step, that of miniaturization (or nanification, as miniaturization down to the nanoscale is called)—see Figure 1.1. As with the case of direct substitution, the advantages should be easy to describe and the consequences easy to predict, even if an actual demonstration is likely to be more difficult to achieve.

A curious, but apparently quite common difficulty encountered by highly upstream nanotechnology suppliers is related to the paradox (attributed to Jean Buridan) illustrated by an ass placed equidistantly between two equally attractive piles of food, hence unable to decide which one to eat first, and which starved to death through inaction. Potential buyers of nanoparticles have complained that manufacturers tell them "we can make any kind of nanoparticle for you." This is unhelpful for many downstream clients, because their knowledge of nanotechnology might be very rudimentary, and they actually need advice on the specification of which nanoparticles will enhance their product range. Start-up companies that offer a very broad product range typically are far less successful than those that have focused on one narrow application, despite the allure of universality (see Chapter 13), not least because—ostensibly—it widens the potential market.

On the other hand, for a larger company universal technologies are attractive commercial propositions. They allow flexibility to pursue alternative market applications, risks can be diversified, and research and development costs can be amortized across separate applications. The variety of markets is likely to have a corresponding variety of stages of maturity, hence providing revenue opportunities in the short, medium, and long terms. As commercialization develops, progress in the different applications can be compared, allowing more objective assessments of performance than in the case of a single application; and the breadth and scope of opportunity might attract more investment than otherwise [4].

10.2 The radical nature of nanotechnology

But nanotechnology is above all a radical, disruptive technology whose adoption implies discontinuity with the past. In other words, we anticipate a qualitative difference between it and preceding technologies. In some cases, this implies a wholly new product; and at the other extreme an initially quantitative difference (progressive miniaturization) may ultimately become qualitative. While a generic technology has breadth, a radical technology has depth, since changes, notably redesign, might be needed all the way down the supply chain to the consumer; they affect the whole of the supply chain, whereas an incremental technology typically only affects its immediate surroundings. Insofar as the very definition of nanotechnology includes words such as "novel" and "unique" (see Section 1.6), "true" nanotechnology can scarcely be called anything but radical, otherwise it would not be nanotechnology [5].

The costs of commercialization are correspondingly very high. Redesign at a downstream position is expensive enough, but if it is required all the way, the costs of introduction might be prohibitive. Furthermore, the more radical the technology, the greater the uncertainty in predicting the market for the product. High uncertainty is equivalent to high financial risk, and the cost of procuring the finance is correspondingly high. "Costly" might mean simply that a high rate of interest is payable on borrowings, or it might mean that capital is difficult to come by at all. This is in stark contrast to an incremental technology, for which the (much smaller) amount of capital required should be straightforward to procure, because the return on investment should be highly predictable.

In addition, the more radical the innovation, the more likely it is that other innovations will have had to be developed in parallel to enable the one under consideration to be exploited. If these others are also radical, then maybe there will be some favorable synergies, since comprehensive redesign is anyway required even for one. There may also be regulatory issues, but at present nanotechnology occupies a rather favorable situation, because there is a general consensus among the state bureaucracies that manage regulation that nanoparticulate X, where X is a well known commercially available chemical, is covered by existing regulations governing the use of X in general. This situation stands in sharp contrast to the bodies (such as the FDA in the USA) entrusted with granting the *nihil obstat* to new medicinal drugs, which, following the thalidomide and other scandals, have become extremely conservative. Things are, however, likely to change, because one of the few clearly articulated recommendations of the influential Royal Society of London–Royal Academy of Engineering report on nanotechnology [6] was that the biological effects of nanoparticles require more careful study before allowing their widespread introduction into the supply chain (cf. Section 5.6).

The implications go even further, because an existing firm's competences may be wholly inadequate to deal with the novelty. Hence, the infrastructure required to handle it will include the availability of new staff qualified for the technology, or the possibility of new training for existing staff.

Nanotechnology is clearly both radical and universal. This combination is in itself unusual, and justifies the need to treat nanotechnology separately from other technically based sectors of the economy.

10.3 Intellectual needs

As well as material capital, the innovating company also has significant intellectual needs. It is perhaps important to emphasize the depth of those needs. Although the scientific literature today is comprehensive and almost universally accessible, simply buying and reading all the journals would not unlock the key to new technology: one needs to be an active player in the field just to understand the literature, and one needs to be an active contributor to establish credibility and allow one to participate in meaningful discussions with the protagonists [7].

Science, technology, and innovation all require curiosity, imagination, creativity, an adventurous spirit, and openness to new things. Progress in advanced science and technology requires years of prior study in order to reach the open frontier, and to perceive unexplored zones beyond which the frontier has already passed. Governments, mindful that innovation is the wellspring of future wealth, do their best to foster an environment conducive to the advance of knowledge. Hence it is not surprising that the state typically plays a leading role in the establishment of research institutes and universities.

Nevertheless, in this "soft" area of human endeavor it is easy for things to go awry. The linear Baconian model has recently recaptured the interest of governments, who wish to expand the controlled legal framework supposedly fostering commercially successful innovations (such as the system of granting patents) by extending their control upstream to the work of scientists. Even in the Soviet Union under Stalin, a world steeped in state control, it was realized that extending it that far was inimical to the success of enterprises (such as the development of atomic weapons) that were considered to be vital to the survival of the state.

This lesson seems to have been forgotten in recent decades. The system of allocating blocks of funds to universities every 5 years or so and letting them decide on their research priorities has been replaced by an apparatus of research councils to which scientists must propose projects, for which funds will be allocated if they are approved. Hence, the ultimate decision on what is important to investigate is taken away from the scientists themselves and put in the hands of bureaucrats (some of whom, indeed, are themselves former scientists, but obviously cannot maintain an acute knowledge of the cutting edge of knowledge). To any bureaucrat, especially one acting in the public interest, the file becomes the ultimate object of importance (for, as C.N. Parkinson points out [8], there may subsequently be an inquiry about a decision, and the bureaucrat will be called upon to justify it). Therefore great weight is placed on clearly measurable outcomes ("deliverables") of the research, which should be described in great detail in the proposal, so that even an accountant would have no difficulty at the end of the project in ascertaining whether they had indeed been delivered. The most common criticism of proposals by reviewers seems to be that they lack sufficient detail, a criticism that is frequently fatal to the chances of the work being funded. Naturally, such an attitude does nothing to encourage adventurous, speculative thinking. Even de Gaulle's Centre National de la Recherche Scientifique (CNRS), modeled on the Soviet system of Academy institutes, and offering a place where scientists can work relatively free of constraints, is now in danger of receiving a final, mortal blow (in fact, for years the spirit of the endeavor had not been respected; the resources available to scientists not associated with any particular project had become so minimal that they were only suitable for theoretical work requiring neither assistants nor apparatus).

One can hardly imagine that such a system could have been introduced, despite these generally recognized weaknesses, were there not failings in the alternative system. Indeed we must recognize that the system of allocating a block grant to an institute only works under conditions of "benign dictatorship." Outstanding directors of institutes (such as the late A.M. Prokhorov, former director of the General Physics Institute of the USSR Academy of Sciences) [9] impartially allocated the available funds to good science—"good" implying both intellectually challenging and strategically significant. Unfortunately, the temptations to partiality are all too frequently succumbed to, and the results from that system are then usually disastrous. A possible alternative is democracy: the faculty of science receives a block grant, and the members of the faculty must agree how to divide it among themselves. It is perhaps an inevitable reflection of human nature that this process almost invariably degenerates into squabbling. Besides, the democratic rule of simple majority would ensure that the largest blocs appropriated all the funds. Hence in order for democracy to yield satisfactory results, it has to be accompanied by so many checks and balances it ends up being unworkably cumbersome.

Is there a practical solution? Benign dictatorship would appear to yield the best results, but depends on having an inerrant procedure for choosing the dictator; in the absence of such a procedure (and there appear to be none that are socially acceptable today) that way has to be abandoned. The opposite extreme is to give individual scientists a grant according to their academic rank and track record (measured, for example, by publications). This system has a great deal to commend it (and, encouragingly, appears to be what the Research Directorate of the European Commission is aiming at with its recently introduced European Research Council, awarding research grants to individual scientists [10]). A weakness is that, almost inevitably, scientists work in institutes, with all that implies in terms of possibilities for partiality in the allocation of rooms and other institutional resources by those in charge of the administration, who are not necessarily involved in the actual research work.

A more radical proposal is that of Selfridge—calling for half-baked ideas, from which a national commission would select some for funding, not so much on the basis of obvious merit (for those should be picked up anyway by private enterprise), but more on the basis of fancy [11]. One would have thought that the potential benefits were sufficiently large that it would be worth allocating at least 10% of the regular funding budget to a trial of Selfridge's idea.

10.4 Company–university collaboration

The greatest need seems to be to better align companies with university researchers. Many universities now have technology transfer offices, which seem to think that great efforts are needed to get scientists interested in industrial problems. In reality, however, rarely are such efforts required—a majority of devoted scientists would agree with A.M. Prokhorov about the impossibility of separating basic research from applied (indeed, these very expressions are really superfluous) [12]. On the contrary, university scientists are usually highly interested in working with industrial colleagues; it is typically the institutional environment that hinders them from doing so more effectively. Somehow an intermediate path needs to be found between the consultancy (which is often too detached and less effective for the company than access to the available expertise would suggest should be the case), the leave of absence of a company scientist spent in a university department (which seems to rapidly detach the researcher from "real-life" problems), and the full-time company researcher, who in a small company may be too preoccupied with daily problems that need urgent attention, or who in a larger company might be caught up in a ponderous bureaucracy. Furthermore, companies tend to be so reticent about their real problems that it may be impossible for an external scientist to make any useful contribution to solving them. One seemingly successful model, now being tried in a few places, is to appoint company "researchers in residence" in university departments—they become effectively members of the department, but would be expected to divide their time roughly equally between company and university. Such schemes might be more effective if there were a reciprocal number of residencies of university researchers in the company. Any expenses associated with these exchanges should be borne by the company, since it is they who will be able to gain material profit from them; misunderstanding over this matter is sometimes a stumbling block. It is profoundly regrettable that the current obsession with gaining revenue from the intellectual capital of universities has to some degree poisoned relationships between them and the rest of the world. The free exchange of ideas is thereby rendered impossible. In effect, the university becomes simply another company. If the university is publicly funded, then it seems right to expect that its intellectual capital should be freely available to the nation funding it. In practice, "nation" cannot be interpreted too literally; it would be contrary to the global spirit of our age to distinguish between nationals and foreigners—whether they be students or staff—and if they are roughly in balance, as they typically are, there should be no need to do so.

A further danger attending the growth of company–university collaboration is the possibility of intellectual corruption rooted in venality [13]. It seems unlikely that this can be prevented by statute; only by rigorous personal integrity. It is, of course, helpful if the relevant institutional environment favors it rather than the opposite.

10.5 Clusters

Evidence for the importance of personal intellectual exchanges comes from the popularity, and success, of clusters (also called innovation hubs) of high-technology companies that have nucleated and grown, typically around important intellectual centers, such as the original Silicon Valley in California, the Cambridges of England and Massachusetts, and the Rhône-Alpes region of south-eastern France [14].

An additional feature of importance for radical nanotechnology is the availability of centralized fabrication and metrology facilities, the use of which solely by any individual member of the cluster would scarcely be at a level sufficient to justify the expense of installing and maintaining them.

Some clusters are initiated and succoured by government support. Examples are the semiconductor, digital display, and notebook PC clusters in Taiwan; the biomedical research cluster in Singapore, the micro-electronics cluster in Grenoble (France), and the life sciences and IT clusters in Shanghai-Pudong.

There is, as yet, no "theory of clusters" [15], but it is fascinating to note that empirical studies have found that larger cities are more innovative (as measured, for example, by patenting activity) than smaller ones [16]. This should provide some guidance for locating clusters when they are planned rather than arising spontaneously. It should be noted that diverse clusters are likely to be more successful in generating innovation than specialized ones [17].

10.6 Assessing demand for nanotechnology

When a decision has to be made regarding the viability of investment in a nanotechnology venture, the costs can usually be fairly well predicted. In more traditional industries, these costs might be extremely well determined. Given nanotechnology's closeness to the fundamental science, however, it is quite likely that unforeseen difficulties may arise during the development of a product for which proof of principle has been demonstrated. By the same token, difficulties may have been anticipated on the basis of present knowledge, but subsequent discoveries may enable a significant shortcut to be taken. On balance, these positive and negative factors might compensate each other; it seems, however, to be part of human nature to minimize the costs of undertaking a future venture when the desire to undertake it is high [18]. There is a strong element of human psychology at work here.

The development, innovation, and marketing costs determine the amount of investment required. The return on investment arises through sales of the product (the market); that is, it depends on demand, and the farther downstream the product, the more fickle and unpredictable the consumer.

A starting point for assessing some of these costs would appear to be the elasticities of supply and demand. Extensive compilations have been made in the past, [19] and (updated) might be useful for products of substitution and innovation. Of course, this would represent only a very rudimentary assessment, because all the cross-elasticities would also have to be taken into account. Furthermore, the concept has not been adequately developed to take quality into account (which is sometimes difficult to quantify).

A perpetual difficulty is that only very rarely can the impact of the introduction of a new product be compared with its non-introduction. Change may have occurred in any case and even the most carefully constructed models will usually fail to take into account the intrinsic nonlinearities of the system.

10.6.1 Modeling

A decision whether to invest in a new technology will typically be made on the basis of anticipated returns. While in the case of an incremental technology these returns can generally be estimated by simple extrapolation from the present situation, by definition for any radical (disruptive) technology there is no comparable basis from which to start. Hence one must have recourse to a model, and the reliability will depend upon the reasonableness of the assumptions made. Naturally, as results start to come in from the implementation of the technology, one can compare the predictions of the model with reality and adjust and refine the model. An example of this sort of approach is provided by cellular telephony: the model is that the market consists of the entire population of the Earth.

One of the problems of estimating the impact of nanotechnology tends to be the overoptimism of many forecasters. The "dotcom" bubble of 2000 is already a classic example. Market forecasts for mobile phones had assumed that almost every adult in the world would buy one and it therefore seemed not too daring a leap to assume that they would subsequently want to upgrade to the 3G technology. Although the take-up was significant it was not in line with the forecast growth of the industry—with all too obvious consequences. Nanotechnology market forecasting is still suffering from the same kind of problem; for example, will every young adult in the requisite socio-economic group buy an iPod capable of showing video on a postage stamp-sized screen? The next section offers a more sober way to assess market volume.

10.6.2 Judging innovation value

The life quality index Q, to be discussed in more detail in Section 18.2, is defined as

$$Q = G^q X_{\mathrm{d}}, \tag{10.1}$$

where G is average work-derived annual earnings, q is optimized work–life balance (here defined as $q = w/(1 - w)$, where w is the optimized average fraction of time spent working, and considered as a stable constant with a value $q = 1/7$ for industrialized countries), and X_d is discounted life expectancy. From the manufacturer's viewpoint, any substitutional or incremental innovation that allows specifications to be maintained or surpassed without increasing cost is attractive. But how will a prospective purchaser respond to an enhanced specification available for a premium price? Many such consumer products are now available, especially in Japan [20]. Theoretically, if the innovation allows a chore to be done faster, then its purchase should be attractive if the increase of Q due to the increase of q is more than balanced by the decrease of Q due to the diversion of some income into the more expensive product.

10.6.3 Anticipating benefit

In which sectors can real benefit from nanotechnology be anticipated? What is probably the most detailed existing analysis of the economic consequences of molecular manufacturing assumes blanket adoption in all fields, even food production [21]. Classes of commodity well suited for productive nanosystems (PNs) include those that are intrinsically very small (e.g., integrated electronic circuits) and those in which a high degree of customization significantly enhances the product (e.g., medicinal drugs) [22]. In many other cases (and bear in mind that even the most enthusiastic protagonists do not anticipate PNs to emerge in less than 10 years), a more technologically modest introduction of nanotechnology may allow a familiar product to be upgraded more cheaply than by conventional means.

Any manufacturing activity has a variety of valid reasons for the degree of centralization and concentration most appropriate for any particular type of product and production. The actual degrees exhibited by different sectors at any given epoch result from multilevel historical processes of initiation and acquisition, as well as the spatial structure of the relevant distributions of skills, power, finance, and suppliers. The inertia inherent in a factory building and the web of feeder industries that surround a major center mean that the actual situation may considerably diverge from a rational optimum.

The emergence of a radical new technology such as nanoscale production will lead to new pressures, and opportunities, for spatial redistribution of manufacturing, but responses will differ in different market sectors. They will have different relative advantages and disadvantages as a result of industry-specific changes to economies of scale, together with any natural and historic advantages that underlie the existing pattern of economic activities. But we should be attentive to the possibility that the whole concept of economies of scale will presumably become irrelevant with the advent of productive nanosystems, and will have intermediate degrees of irrelevance corresponding to intermediate stages in the development of nanotechnology.

10.7 Technical and commercial readiness (availability) levels

The notion of technology readiness level (TRL) originated in the NASA Advanced Concepts Office [23]. It was specifically designed to characterize the progress of technologies used in space missions. Despite this association with a very closed and controlled uncommercial environment, it has since then been widely adopted to characterize general industrial technologies, for which it is far from being appropriate. Hodgkinson et al. proposed modifications to the scale to render it more suitable for general use (Table 10.1) [24]. The third column in the table has been added to characterize the nature of the activity culminating in the achievement of each level. "Science" means here "research"; that is, the application of the scientific method to make discoveries and create new knowledge; and "engineering" means "development"; that is, the application of existing knowledge to perfect something in order to enable it to deliver practical, useful results [12]. Of course, many new things are also discovered during development but the work is directed toward a specific goal, whereas science is open-ended. Science can be considered to have become technology by TRL 4.

Even the modified scheme of TRL is, however, inadequate to represent the progression of nanotechnology, because of some unique features, especially its extremely rapid progression and the intermingling of academic and industrial research. Hence, a scheme of *nanotechnology availability levels* (NAL) is proposed to more accurately describe the state of a given part of nanotechnology with respect to the desired application (Table 10.2).

Note that the levels are not ordered in a strict sequential hierarchy throughout: NAL 4 may well accompany NAL 3; availability at NAL 5 may only be of research grade material. Levels 0 to 6a would typically be reached in the research environment;

Table 10.1 Technology or commercial readiness levels (after Mankins *loc. cit.* and Hodgkinson et al. *loc. cit.*). The third column identifies the nature of the work involved in arriving at that level.

TRL	Description	Nature[a]
1	Basic principles observed	S
2	Proof of concept	S
3	Technology application formulated	S/E
4	Component validation in the laboratory environment	S/E
5	Component validation in the real environment	E
6	Prototype system demonstration in the real environment	E
7	Commercial introduction	E
8	Commercial success	E

[a] *Science (S) or engineering (E).*

Table 10.2 Nanotechnology availability levels (NAL), giving next step requirements (NSR).

NAL	Description	NSR[a]
0	Idea	R1
1	Basic principles observed	R1,V
2	Proof of concept	I
3	Technology application(s) formulated	IP
4	Demonstration for some specific application	P
5	Available in some form	St
6a	COTS[b] availability	R3,SC
6b	Validated for a desired application	R3,SC
7a	Complete supply chain established	R4
7b	Incorporated into the product	U,R4
8	History of success	R4

[a] *Key to next step requirements (NSR):*
[b] *Commercial off-the-shelf.*

R1 Exploratory laboratory work
V Verification
I Inspirational thinking
IP Patent application
P Development of a feasible production route
R2 Research to establish whether the technology can be used for the desired application and whether it is superior to existing technology
St Standardization
R3 Testing in the desired application
U Use
SC Establishing the supply chain
R4 Optimization

levels 0–3 might well be carried out in an academic environment; while institutes of technology might continue work through levels 4 and 5, possibly then transferring operations to a spin-off company in order to progress the technology further. NAL 6a corresponds to the interest of the supplier and ultimate manufacturer; NAL 6b and 7b to the interest of the ultimate end-user, which is likely to have to shoulder the burden of the research work labeled R3. 6a and 6b are likely to take place in parallel; likewise for 7a and 7b. Ongoing R4 is generally required for sustainable deployment. Nowadays it is common for academic laboratories to seek to patent new technologies, which introduces some drag into the reporting—a published account of the principles (NAL 1) may only appear after levels 4 and 5 have been reached. Note also the deliberate alignment of the TRL, CRL, and NAL. Therefore, in every case the actual numbers should convey roughly the same meaning.

Level 7a can be further divided into the following maturity levels [25]:

1. Manufacturable solutions are not known.
2. Interim solutions are known.

3. Manufacturable solutions are known.
4. Manufacturable solutions exist and are being optimized.

Level 7b could already begin to be accessed once level 7a.2 or 7a.3 is reached (as we can respectively label items 2 and 3 from the above list of manufacturing maturity levels).

Nanotechnologies that can meet short-term needs for a particular purpose have typically already been developed for some other application (i.e., NAL 3, or 4). Nevertheless, given that the material is available in some form, it is essentially a matter of straightforward engineering development to adapt it for the specific requirement, without excluding the possibility that unexpected problems arise, needing more fundamental investigation of, probably, a very specific aspect. Short-term technologies already have some commercial activity, which may not be in the same area of application as the one of interest to the product developer. Medium-term technologies are being actively pursued in university and other academic laboratories as well as in the research laboratories of leading high technology industries (although activity in the latter is usually only revealed when a patent application is filed). Long-term technologies are currently less actively pursued in academic laboratories than might be imagined (because they are increasingly forced to rely upon short-term research contracts with government funding agencies that require detailed, prespecified outputs); the main thrust is currently from privately funded nonprofit institutions and theoretical work by academics not forced to rely on external funding.

Manufacturability. It should be borne in mind that the ultimate goal of most nanotechnology research is high-volume, low-cost manufacture of materials and devices. It seems that none, or hardly any, of the great number of research papers address this issue. Making single electron devices and the like tend to be a *tour de force* of individual skill and ingenuity, but there is no clear route thence to the higher availability levels. Unfortunately, even the more detailed and better adapted (compared to TRL or CRL) nanotechnology availability levels (NAL) do not capture the possibility of an intrinsic limitation to the new technology that will prevent it ever getting beyond level 3. Manufacturability, therefore, needs careful attention that will probably require customized analysis, since it is not usually addressed in the literature (cf. Section 4.2).

10.8 Predicting development timescales

This is difficult for new technologies because they progress exponentially; the best known example is probably Moore's law, which states that the number of components on a VSLI chip now doubles approximately every 2 years. Hence, if we now have 10^9 components on a chip, we can expect there to be $2^6 \times 10^9$, or almost 10^{11}, components after 9 years [26]. This kind of prediction of exponential development can only be made with some hope of reliability when an entire industry is taken into consideration. The rate of development in a narrow field is much harder to predict. It usually depends strongly on the volume of available resources, especially manpower. Furthermore, discoveries in other, seemingly unrelated, fields may solve bottlenecks and have other dramatic, unpredicted effects [27].

Short-term (<5 years) needs are typically met by substitution of a traditional material or extant device (which might already be micro) by nanotechnology. In simple cases there is a clearly predictable benefit (e.g., less cost for the same performance, or more performance for the same cost). In more complex cases, there is a trade-off; if benefits and disbenefits are size-dependent but the dependences are of opposite sign, there will be a definite characteristic size (crossover point) below which there is no advantage to be gained. This applies, for example, to microelectromechanical systems (MEMS) used, e.g., as inertial sensors such as accelerometers; performance is degraded upon miniaturization down to the nanoscale.

These applications need essentially development, including optimization, of existing technologies. The technologies may have been developed with other purposes in mind, in which case adaptation to the specific need is required. Computer modeling may be brought to bear if experiments are difficult, in order to narrow the scope and reduce the required number of experiments. In all these applications the first step must be to draw up the detailed specifications of what is required. Available knowledge can then be used to select candidate solutions, and the final choice must depend on comparative experiments.

Although at first sight there seems to be a plethora of nanomaterials ripe for short-term applications, beyond niche products (e.g., tennis balls and rackuets, stain-proof cravats) the lack of standardization of raw materials is a definite handicap. Research work, therefore, has to rely on the material produced by a particular supplier, with no guarantee that the results remain valid with materials from other suppliers (even if the material is nominally the same). If the research is successful and the material is shown to be fit for purpose, a similar difficulty arises at the next level, because of the fragility of a supply chain depending on a single manufacturer. That is why, rather than through achieving any specific research results, the introduction of exchange-based trading of commoditized nanomaterials is a prerequisite for sustainable short-term development (cf. Section 11.3).

Medium-term (5–15 years) applications require research (of type R2 according to Table 10.2), both experiments and more or less profound theoretical analysis in order to be able to appraise the practicability of selecting them for development.

The long-term (>15 years) applications of nanotechnology, apart from those listed in the previous section that belong to the far end of the medium term, are essentially associated with the development of nanoscale assemblers (also known as bottom-to-bottom "fabbers," mechanosynthesizers, etc.) as universal fabrication tools, as suggested by R.P. Feynman in his 1959 Caltech lecture and later developed in much more detail by K.E. Drexler, R.A. Freitas Jr, R.C. Merkle, and others. Although these assemblers would therefore appear to constitute the very core of the mainstream of nanotechnology, and although the US National Nanotechnology Initiative (NNI) launched in 2001 makes strong reference to Feynman, that initiative has ended up giving very little support to the development of bottom-to-bottom fabrication, which is, therefore, nowadays carried out in a small number of university

Table 10.3 The infiltration of science into industry (after Bernal).[a]

Stage	Description	Characteristic Feature(s)
1	Increasing the scale of traditional industries	Measurement and standardization
2	Some scientific understanding of the processes (mainly acquired through systematic experimentation in accord with the scientific method)	Enables improvements to be made
3	Formulation of an adequate theory (implying full understanding of the processes)	Possibility of completely controlling the processes
4	Complete integration of science and industry, extensive knowledge of the fundamental nature of the processes	Entirely new processes can be devised to achieve desired ends

[a] *Loc. cit.*

or private institutions such as the nonprofit Institute for Molecular Manufacturing in Palo Alto. A similar situation prevails in Europe. The situation in China is not reliably known.

The Nano Revolution will consummate the trend of science infiltrating industry that began with the Industrial Revolution and which can be roughly described in four stages of increasing complexity (Table 10.3) [28]. Note that Stage 4 also encompasses the cases of purely scientific discoveries (e.g., electricity) being turned to industrial use. Clearly nanotechnology belongs to Stage 4, at least in its aspirations; indeed nanotechnology is the consummation of Stage 4; a corollary is that nanotechnology should enable science to be applied at the level of Stage 4 to even those very complicated industries that are associated with the most basic needs of mankind, namely food and health. Traditional or conventional technologies (as we can label everything that is not nanotechnology) also have Stage 4 as their goal but in most cases are still far from realizing it.

10.9 Nanometrology

It might well be asserted that metrology invented nanotechnology. Apart from the nanoparticles long synthesized by chemists, the first practical development in nanotechnology was the scanning tunneling microscope, followed by the atomic force microscope and there is still a growing number of other scanning probe microscopies. These instruments were able to resolve height differences with subatomic resolution, and their lateral resolution is now good enough to enable individual atoms to be imaged. Moving atoms around on a solid surface, which can readily be accomplished by an atomic force microscope, is a prototype of the process of atom-by-atom assembly that will, it is hoped, form the basis of productive nanosystems.

Metrology has long been associated with the obligatory use of standard weights and measures. England's Magna Carta (1215) makes reference to such an obligation, the history of which goes back to the earliest civilizations such as Egypt. Clearly such obligations cannot be enforced unless there are reliable methods for quantifying masses and lengths, as well as the other basic quantities such as electric current, temperature, luminous intensity, amount of substance, and (sidereal) time that constitute the Système International d'Unités; all the other units can be derived from these.

Owing to the intrinsic difficulties of making measurements at the nanoscale, many nanotechnologists end up having to follow an empirical, trial-and-error approach to product development, instead of a systematic and controlled one. Although the former may require little additional capital expenditure, it is lengthy and, hence, expensive, apart from the significant material waste that generally ensues. Presently, this appears to be very much the case in nanomedicine. The developers of diagnostic nanodots and biosensors lack the metrology tools that they would need to improve device performance in a rational and controlled manner. The lack of such tools not only impedes development, but is also an obstacle to formulating procedures for measurement, quality control, and reproducibility of manufacture. The deficiency in nanometrology tools is particularly glaring in technologies involving the life sciences, although there has been encouraging recent progress in developing tools based on interrogating the evanescent optical fields created at the surface of planar optical waveguides functioning as substrata for living cells and biomolecular thin films.

A similar deficiency can be seen in the development of nanodevices for energy conversion and storage. While purely structural nanometrology has become well developed, thanks, above all, to techniques like the atomic force microscope and electron microscopy, together with auxiliary techniques such as fast ion bombardment (FIB) that are extremely useful for sample preparation, device development can still be very slow because of the lack of functional microscopies. For example, what would be needed for the development of novel solar cells is a technique able to simultaneously characterize the three-dimensional domain structure and the optoelectronic properties under device operating conditions. Nevertheless, the seemingly endless versatility of the scanning probe microscope may come to the rescue, since the scanning tip can be made to be sensitive to photoconductivity. Another example is the scanning Kelvin probe microscope (SKPM) for the study of transparent conducting nanocomposites (to replace indium tin oxide). The SKPM can measure both surface topography and work function (i.e., the ease of extracting electronic charge from a material) [29]. Other extensions of the scanning probe concept include scanning electrochemical microscopy (SECM), which can be usefully applied to develop novel materials for fuel cells [30]; and scanning ion current microscopy (SICM), which allows the topography of living cells to be mapped with less perturbation of the cell than with conventional atomic force microscopy; and electrostatic force microscopy [31], which is very useful for precise measurements of the thickness of oligolayer graphene.

Looking further into the future, the development of quantum cryptography based on encoding information into the quantum states of single photons will also require a metrology framework for independently assessing performance. Challenges of this magnitude are generally only achievable through multinational collaboration.

Whereas in the past metrology has been firmly embedded in the domain of physics, the introduction of nano-objects, which can quickly be dispersed into the environment in enormous numbers with unknown and possibly harmful effects, has prompted the development of nano-exotoxicology as a branch of physics rather than remaining within the traditional domain of naturalists. A key development here is the creation of reference materials for calibrating instruments and optimizing protocols for sample preparation and testing. Such reference materials are no less important in this domain than they are for the length calibration of scanning probe microscopes.

The interaction between metrology and nanotechnology operates in both directions. Recently graphene, the archetypical nanomaterial, was used to effect the most precise measurements of the quantum Hall effect ever achieved [32], which will assist the work to redefine the fundamental units of mass and electric current.

Furthermore, the scanning tunneling and atomic force microscopes, which have played such a central role in the development of nanotechnology, are the preferred tools for the development of the assemblers envisioned by Richard Feynman for the fabrication of devices and materials to atomic specifications, and given more concrete form by the subsequent work of Eric Drexler. With these instruments single atoms can be picked up from, displaced along, and deposited on a substrate.

Despite all the developments, there is still no single instrument that can fulfil all the needs of nanotechnologists. Figure 10.2 summarizes the most important available techniques for topographic and chemical analysis, plotted according to their spacial and chemical resolution.

10.10 Standardization of nanotechnology

It has long been recognized that the adoption of standards has a strongly positive effect on commercial life, so much so that, historically, adopting them has often been a legally enforced obligation. A straightforward example is offered by railways. A company fixing a unique gauge for its tracks retains a complete monopoly over their exploitation but the impossibility of running trains through to other networks creates a net disbenefit. Occasionally there were strategic reasons for preventing such through running (as in the case of Russia), especially important as a time when the military potential of railways had begun to be perceived, but nowadays the nuisance of having to change trains, or change the bogies, at the border is only mitigated by Russia being such a vast country; border traffic only constitutes a minute proportion of the whole.

Standards can be normative, specifying what shall be done (e.g., a specific test method) or simply informative. For the most part their adoption is nowadays voluntary. The fact that they have been drawn up through a careful process of achieving consensus, preferably international (the leading organization is the ISO) is an obvious encouragement for their adoption. Voluntary submission to an engineering constraint removes constraints to trade [33].

The first stage of standardization is to agree on the definitions of words. Thus, the first publications of the ISO Technical Committee (TC) 229—Nanotechnologies

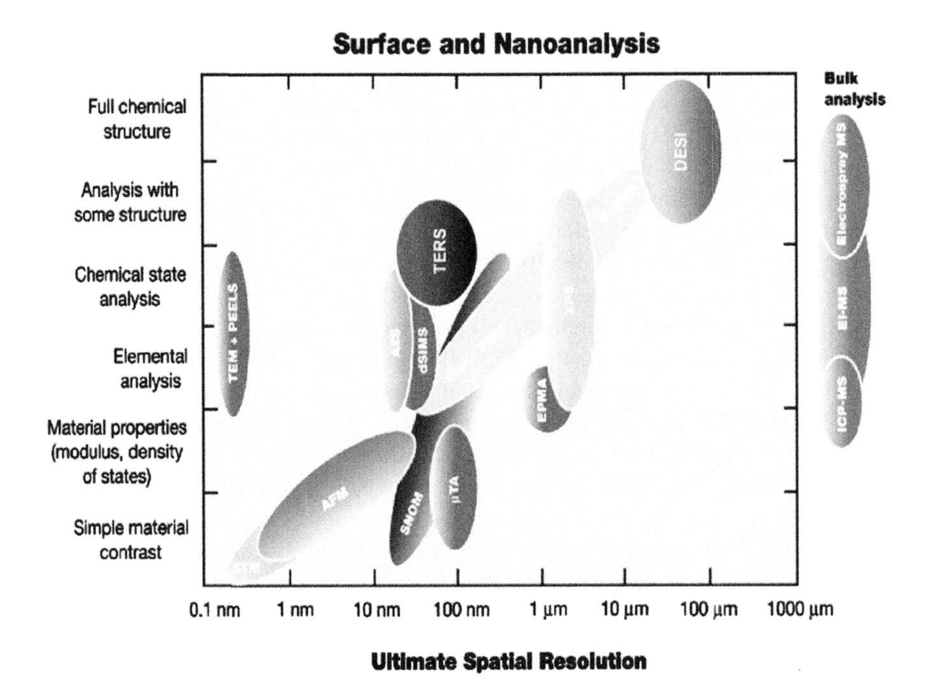

FIGURE 10.2

Techniques for surface and nanoanalysis. AES, Auger electron spectroscopy; AFM, atomic force microscopy; DESI, desorption electrospray ionization; dSIMS, dynamic SIMS (secondary ion mass spectrometry); EPMA, electron probe microanalysis; G-SIMS, gentle SIMS; μTA, micro thermal analysis; SNOM, scanning near-field optical microscopy; STM, scanning tunneling microscopy; TERS, tip-enhanced Raman spectroscopy; TEM, transmission electron microscopy; PEELS, parallel electron energy-loss spectrometry; XPS, X-ray photoelectron spectroscopy. Bulk analysis terms: EI-MS, electron ionization mass spectrometry; ICP, inductively-coupled plasma. Figure reproduced from C. Minelli and C.A. Clifford, The role of metrology and the UK National Physical Laboratory in nanotechnology. *Nanotechnol. Perceptions* 8 (2012) 59–75, with permission of Collegium Basilea.

(constituted in 2005) have been vocabularies [34]. The work of other ISO technical committees also impinges on nanotechnology: for example TC 24 (particle characterization); TC 201 (surface chemical analysis); TC 202 (microbeam analysis); and TC 213 (the geometry of surfaces).

10.11 Patents

The patent system has always been an anomaly in the free enterprise economy. It amounts to inventors being accorded monopoly privileges enshrined in law. England appears to have the oldest continuous patenting system in the world, starting with the patent granted in 1449 to John of Utynam for making stained glass. Already by 1623, however, the *Statute of Monopolies* placed clear limitations on the extent of monopoly: both temporal (at that time, a maximum of 14 years) and stipulating that the public interest must be respected.

The patent system greatly expanded with the onset of the Industrial Revolution, the number of patents roughly following the increase in national wealth. Arguments for and against them have continued to this day. The main argument in favor is that patents provide an incentive to the innovator. This was always considered to be rather specious in the case of the individual inventor, but at company level it is used, for example by the pharmaceutical industry. Its argument is that were it not for the period of guaranteed monopoly, it would be difficult to recoup the tremendous costs of research and even more so of development to produce new medicinal drugs, because other companies could simply copy and sell the drugs themselves once they had been placed on the market. Although this view is widespread, it does not have empirical support [35]. It is also rather difficult to see the logic of the argument because exactly the same premises are widely used to justify the privatization of state monopolies in order to open up their services to a multiplicity of competing companies—a guaranteed monopoly is believed to promote inefficiency in the industry to which it applies, and exactly the same appears to be true of the pharmaceutical industry.

The main argument, however, against patents is that they actually stifle innovation. I.K. Brunel was a noted critic on those grounds, and refused to protect any of his own ideas [36]. Telecommunications have been a rich field (and perhaps still are) for such stifling; for example, in 1937 the US Federal Communications Commission declared that the Bell Telephone System had suppressed over 3000 unused patents in order to forestall competition. Many of these patents did not arise through work done within the company, but were acquired and concerned alternative devices and methods for which the company had no need. It has been estimated that 95% of patents are obstructive. A counterargument, albeit somewhat contrived, to the apparent disbenefit is that a great deal of ingenious research must then be done in order to circumvent the web of existing patents.

Simply looking at the history of patent rights (Table 10.4) in different countries, and the contemporaneous progression of their pharmaceutical industries, also renders untenable the notion that patents are essential to incentivize research. If patents were indeed essential for the success of the pharmaceutical industry, most drugs should have been invented and produced in the UK and the USA. In contrast, Germany, Italy, and Switzerland have historically been the leading countries (in 1978, when Italy changed its patent laws, its pharmaceutical industry was the fifth largest in the world) and in World War I the USA had to import dyes from Germany. In France, the repeal of the 1959 ban virtually killed its chemical industry, which migrated to Switzerland (cf. the movie industry migrating to Hollywood to avoid Edison's patents). In the absence of patents, one notices intense innovative activity leading to the constant improvement of productivity. Furthermore, knowledge is not built from discrete, isolated entities; the

Table 10.4 Patent laws in different countries.

Country	Year	State of the Laws
France	1959	The drug as a product NOT patentable
Ditto	1978	1959 ban lifted
Germany	1967	Prior to that date only processes were patentable
Italy	1978	Products patentable
Spain	1986[a]	Products patentable
Switzerland	1907	The process to produce a drug patentable
Ditto	1954	1907 law strengthened
Ditto	1977	Products patentable
UK	1449	Processes and products patentable
USA	1790	The drug itself (the product) patentable
Ditto	1790	The process to produce a drug patentable

[a] *Upon entry to the European Union.*

invention of new products and processes builds on existing ones [37]; the progression has strong internal correlations.

Open source software (cf. Chapter 15) obviously represents a fundamental challenge to the patent system. Open source hardware is now beginning to be explored. Ultimately, nanotechnology will make the two practically indistinguishable from each other.

References and notes

[1] Synonyms for "universal" in this context are "generic," "general purpose," and "platform."

[2] There is a certain ambiguity here, since the nanoscale processors (which could now be called nanoprocessors) have only been introduced very recently. Hence, the majority of extant information processors probably still belong to microtechnology rather than nanotechnology, but the balance is inexorably tipping in favor of nanotechnology.

[3] This can be considered as quasi-axiomatic. It seems to apply to a very broad range of situations. For example, in agriculture the primary grower usually obtains the smallest profit. The explanation might be quite simple: it is customary and acceptable for each purveyor to retain a certain percentage of the selling price as profit; hence, since value is cumulatively added as one moves down the supply chain, the absolute value of the profit will inevitably increase. In many cases the percentage actually increases as well, on the grounds that demand from the fickle consumer fluctuates, and a high percentage profit compensates for the high risk of being left with unsold stock. As one moves upstream, these fluctuations are dampened and hence the percentage diminishes.

[4] Shane E. Academic entrepreneurship. Cheltenham: Edward Elgar; 2004.

[5] In practice, however, some parts of nanotechnology consist of products of simple substitution (cf. Section 5.4).

[6] *Nanoscience and Nanotechnologies: Opportunities and Uncertainties.* London (2004). This conclusion created a considerable stir and triggered a flurry of government-sponsored research projects. Nevertheless, given the considerable literature that already

existed on the harmful effects of small particles—e.g., Revell PA, The biological effects of nanoparticles. Nanotechnol Perceptions 2006;2:283–98 and the many references therein—and the already widespread knowledge of the extreme toxicity of long asbestos fibers, the sudden impact of that report is somewhat mystifying.
[7] Kealey T. Sex, science and profits. London: Heinemann; 2008.
[8] Parkinson CN. In-laws and outlaws. London: John Murray; 1964. p. 134–5.
[9] The author spent some weeks in his institute in 1991. For a published account, see I.A. Shcherbakov, 25 years of the A.M. Prokhorov General Physics Institute of the Russian Academy of Science (RAS). Quantum Electronics 1991;37(2007):895–6.
[10] Unfortunately the procedure for applying for these grants is unacceptably bureaucratic and thus vitiates what would otherwise be the benefit of the scheme; furthermore, the success rate in the first round was only a few percent, implying an unacceptable level of wasted effort in applying for the grants and evaluating them. The main mistake seems to have been that the eligibility criteria were set too leniently. This would also account for the low success rate. Ideally the criteria should be such that every applicant fulfilling them is successful. Incidentally, this criticism is just one of many directed at European Union research funding: a declaration launched in February, 2010 in Vienna entitled "Trust Researchers" attracted more than 13,000 signatures.
[11] Selfridge OG. A splendid national investment. In: Good IJ, editor. The scientist speculates. London: Heinemann; 1962. p. 31.
[12] See also Ramsden JJ. The differences between engineering and science. Measurement and Control 2012;45:145–6.
[13] Ramsden JJ. Integrity, administration and reliable research. Oxford Magazine, Noughth Week, Trinity Term; 2012. p. 6–8; Ramsden JJ. The independence of university research. Nanotechnol Perceptions 2012;8:87–90.
[14] See Breschi S. Knowledge spillovers and local innovation systems: a critical survey. Liuc Papers no 84, Serie Economia e Impresa (27 March 2001) for an assessment.
[15] See, however, Sedgley N, Elmslie B. Do we still need cities? Evidence on rates of innovation from count data models of metropolitan statistical area patents. Am J Econ Sociol 2011;70:86–108.
[16] Carlino GA. Knowledge spillovers: cities' role in the new economy. Bank of Philadelphia Business Rev 2001;4:17–26.
[17] Feldman MP, Audretsch DB. Innovation in cities: science-based diversity, specialization and localized competition. Eur Economic Rev 1999;43:409–29.
[18] This state of affairs has led to the failure of many (geographical) exploratory expeditions. It is understandable, given the prudence (some would say meanness) of those from whom resources are being solicited, but is paradoxical because the success of the venture is thereby jeopardized by being undertaken with inadequate means. Failure might also decrease the chances of gathering support for the future expeditions of a similar nature.
[19] E.g., Houthakker HS, Taylor LD. Consumer demand in the United States: analyses and projections. Cambridge, MA: Harvard University Press; 1970.
[20] Unfortunately in Europe there is still a strong tendency to buy the cheapest, regardless of quality, which of course militates against technological advance.
[21] Freitas Jr RA. Economic impact of the personal nanofactory. Nanotechnol Perceptions 2006;2:111–26.
[22] As far as nanotechnology is concerned, the task of deciding whether agile manufacturing is appropriate is made more difficult by the fact that many nanotechnology products are

available only from what are essentially research laboratories, and the price at which they are offered for sale is rather arbitrary; in other words, there is no properly functioning market. This situation will, however, evolve favorably if the industry embraces the exchange system for trade (Section 11.3).

[23] Mankins JC. Technology readiness levels. NASA Advanced Concepts Office; 6 April 1995.

[24] Hodgkinson J et al. Gas sensors 2. The markets and challenges. Nanotechnol Perceptions 2009;5:83–107. This modified scheme might be called one of commercial readiness levels (CRL), although the new term is not yet in general use.

[25] Taken from the *International Technology Roadmap for Semiconductors* (ITRS).

[26] That does not imply that performance is necessarily 100 times better, because programming abilities to exploit the increased number of components are likely to lag behind.

[27] In industries with strong regulatory controls, such as pharmaceuticals, the need to pass extensive and exhausting tests before commercialization means that any drug, whether involving nanotechnology or not, may take as long as 15 years to reach the market.

[28] Bernal JD. The social function of science. London: Routledge; 1939.

[29] Cuenat A et al. Quantitative nanoscale surface voltage measurement on organic semiconductor blends. Nanotechnology 2012;23:045703.

[30] Nicholson R et al. Electrocatalytic activity mapping of model fuel cell catalyst films using scanning electrochemical microscopy. Electrochimica Acta 2009;54:4525–33.

[31] Burnett T et al. Mapping of local electrical properties in epitaxial graphene using electrostatic force microscopy. Nano Lett 2011;11:2324–8.

[32] Tzalenchuck A et al. Towards a quantum resistance standard based on epitaxial graphene. Nature Nanotechnol 2010;5:186–9.

[33] One can see a similar effect operating in Switzerland, and in some other countries, which have voluntarily decided to adopt certain European Union laws to facilitate trade between their country and the EU. The only negative aspect of standardization is that it limits diversity. In the engineering world new ideas are, however, constantly emerging from individuals and research institutes, ensuring the constant influx of diversity. There is, however, danger in standardization in the living world. For example, Turkey has voluntarily replaced almost all of its indigenous varieties of tomatoes with varieties approved by the European Union, to the detriment of taste and, very likely, nutritional quality. Although the displaced varieties can be preserved in seed banks, in effect they have become extinct and experience shows how difficult it is to recreate lost diversity in agriculture.

[34] The ISO/TS 80004 multipart series.

[35] Mansfield E et al. Imitation costs and patents: an empirical study. Economic J 1981;91:907–18; this study found that the average costs of copying approached 70% of the costs of the original invention.

[36] Rolt LTC. Isambard Kingdom Brunel. London: Longmans, Green & Co.; 1957. p. 217.

[37] E.g., Thursby J, Thursby M. Where is the new science in corporate R & D? Science 2006;314:1547–8.

Further reading

[1] Boldrin M, Levine DK. Against intellectual monopoly. Cambridge: University Press; 2008.

[2] Hatto P. Standardization for nanotechnology. Nanotechnol Perceptions 2007;3:123–30.
[3] Hawkes PW. From fluorescent patch to picoscopy, one strand in the history of the electron. Nanotechnol Perceptions 2011;7:3–20.
[4] Leach RK. Fundamental principles of engineering nanometrology. Amsterdam: Elsevier; 2010.
[5] Ramsden JJ, Horvath R. Optical biosensors for cell adhesion. J Receptors Signal Transduction 2009;29:211–23. Describes techniques for analyzing the evanescent optical fields created at the surface of planar optical waveguides functioning as substrata for living cells and biomolecular thin films, as an alternative to the traditional methods of optical microscopy for investigating biological objects.

CHAPTER

The Fiscal Environment of Nanotechnology

11

CHAPTER OUTLINE HEAD

11.1 Sources of Funds 137
11.2 Government Funding 141
11.3 Endogenous Funding 145
11.4 Geographical Differences between Nanotechnology Funding 148

Figure 11.1 summarizes the overall path of value creation by a nanotechnology company. We need only consider the two most typical types of nanotechnology company: (1) a very large company that is well able to undertake the developments using internal resources; and (2) the very small university spin-out company that in its own special field may have even better intellectual resources than the large company, but which is cash-strapped. Examples of (1) include IBM (e.g., the "Millipede" mass data storage technology) [1] and Hewlett-Packard ("Atomic Resolution Storage" (ARS) and medical nanobots). Examples of (2) are given in Chapter 13.

We have already mentioned Thomas Alva Edison's "1% inspiration, 99% perspiration" dictum. If the research work needed to establish proof of principle costs one monetary unit, then the development costs to make a working prototype are typically 10 units, and the costs of innovation—introducing a commercial product—are 100 units. The last figure is conservative. An actual example is DuPont's introduction of Kevlar fiber: laboratory research cost $6 million, pilot plant development cost $32 million, commercial plant construction cost approximately $300 million, and setting up marketing, sales, and distribution cost $150 million [2]. Moreover, commercial development is typically lengthy. It took about 17 years for Kevlar to reach 50% peak annual sales volume, which was in fact rather fast in comparison with other similar products (31 years for Teflon, 34 years for carbon fibers, and 37 years for polypropylene) [3]. Hence immense sources of capital are necessary; even a large firm may balk at the cost.

11.1 Sources of funds

Four main sources of capital are available: (i) internal funds of the company; (ii) forward selling products; (iii) private investors (mainly angel investment, venture

Applied Nanotechnology, Second Edition. http://dx.doi.org/10.1016/B978-1-4557-3189-3.00011-7

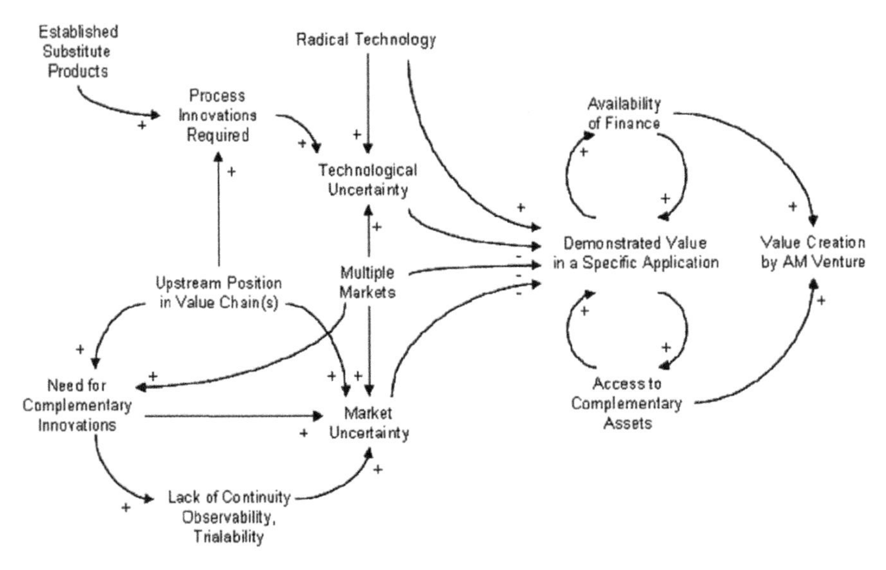

FIGURE 11.1

Diagram of immediate effects for a nanotechnology company. "+" indicates that the factor causes an increase and "−" that it causes a decrease. Reproduced from S. Lubik and E. Garnsey, Commercializing nanotechnology innovations from university spin-out companies. *Nanotechnol. Perceptions* 4 (2008) 225–238 with permission of Collegium Basilea.

capital, and private equity); and (iv) government funds. Very important elements of Figure 11.1 are the two small loops on the right-hand side of the diagram, which are expanded in Figure 11.2.

Generally speaking, (i) is only an option for very large firms, and even they seem to prefer to reserve their cash for acquiring small companies with desirable know-how, rather than developing it themselves. For various reasons connected with problems of internal organization and its evolution, large-company research is often (but not, of course, always) inefficient; the problem is that all firms, as they grow, inevitably also proceed along the road to injelititis [4].

The operation of (ii) is centered on commercial exchanges, such as the Royal Exchange in London (founded in 1565 by Gresham), a similar exchange in Manchester, founded in 1729 to serve the needs of the textile industry, the Baltic Exchange in London, dating from 1744, the Chicago Mercantile Exchange (1874), and so forth, which facilitated not only trade but production wherever investment was needed in advance of the actual delivery of goods (see Section 11.3). Crucially, there must be a prior demand for the goods for the system to work.

While option (iii) has been hitherto probably the main source of funding for fledgling nanotechnology companies, it has notable disadvantages. For one thing it is chaotically organized, often depending on a chance meeting between the company

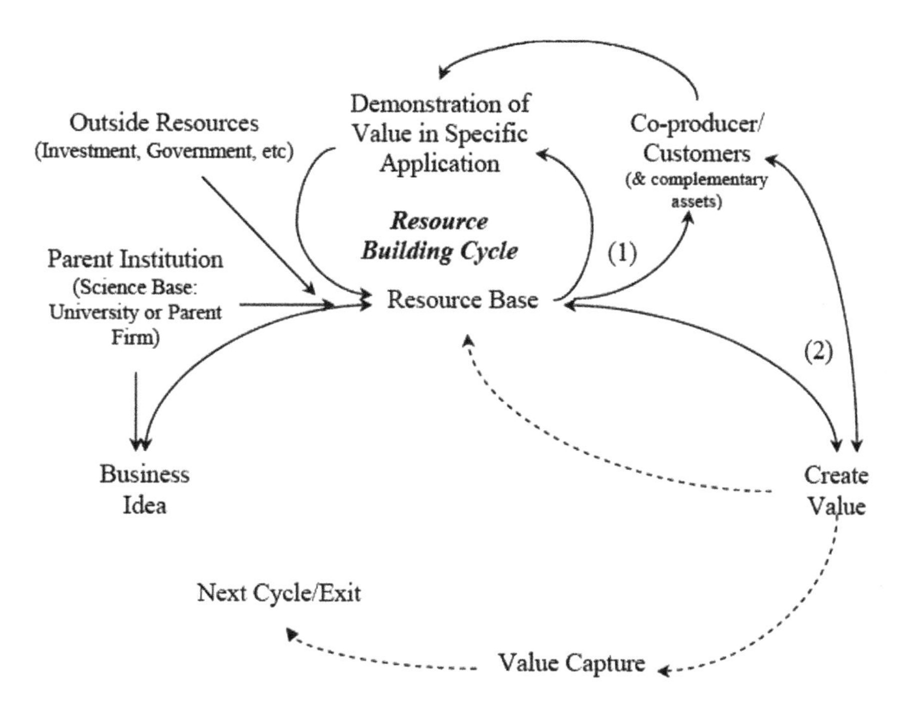

FIGURE 11.2

Diagram of immediate effects for a nanotechnology company, from a slightly different viewpoint compared with Figure 11.1, here focusing on the resource-building cycle. Reproduced from S. Lubik and E. Garnsey, Commercializing nanotechnology innovations from university spin-out companies. *Nanotechnol. Perceptions* 4 (2008) 225–238 with permission of Collegium Basilea.

directors and the investors, and their decision to proceed depends to a large extent on personal factors rather than an objective appraisal of the business. There appears to be an element of luck in finding investors. A social setting (which might be as unpretentious as a college bar) in which would-be investors and technologists mix informally is probably a crucial ingredient. Therefore, it does not seem to be an "efficient" method of financing. More importantly, investment is inevitably accompanied by a loss of control. The priorities of the investors are different from those of the company founders, and the desire of the former to extract the maximum monetary value from their investment as soon as possible sometimes militates against building up the competitive strength of the company's core technology. Venture capitalists usually prefer not to commoditize the products of the companies in which they invest, because they believe that by not doing so they can charge premium prices, but it is against the long-term interest of the industry and inimical to option (ii).

Option (iv) is fraught with difficulties, as will be further discussed in Section 11.2. The establishment of extensive state programs to support nanotechnology research and development is presumably based on the premise that nanotechnology is something emerging from fundamental science, implying that there is insufficient interest from industry willing to lavish funds upon its development. Government largesse might actually hinder development, however. It has long been a criticism of the European Union "Framework" research and technical development programs that they actually hinder innovation in European industry [5]. Generic weaknesses of government funding programs are: excessive bureaucracy, which not only saps a significant proportion of the available funds, but also involves much unpaid work (peer review) by working scientists, inevitably taking time away from their own research; excessive interference in the thematic directions of the work supported, which almost inevitably leads in the wrong direction, since by definition the officials administering the funds have left the world of active research, hence are removed from the cutting edge, nor are they embedded in the world of industrial exigencies; an excessively leisurely timetable for deciding which work to support—12 months is probably a good estimate of the average time that elapses *after* submitting a proposal before the final decision is made by the research council, and to this should be added the time taken to prepare the proposal (9 months would be a reasonable estimate), in which an extraordinary level of detail about the proposed work must typically be supplied (to the extent that, in reality, some of the work must already be done in advance in order to be able to provide the requested detail), and further months elapse after approval before the work can actually begin, occupied in recruiting staff and ordering equipment (6 months would be typical). Operating therefore on a timescale of two or more years between having the idea and actually beginning practical work on testing it, it is little wonder that research council proposals tend to be repositories for incremental, even pedestrian work, the main benefit of which are the accompanying so-called overhead payments that help to maintain the central facilities of the proposer's university.

The above four options are not exhaustive. Another possible route is for the nanotechnology company to enter into a close partnership with a company established in the application area for which the new technology is appropriate. The pooling of complementary interests seems to create a powerful motivation to succeed in the market (see also Chapter 13). Nevertheless, many large companies find it difficult to make a business case for nanomaterials. Typical is the automobile industry's attitude toward advanced composites. It is problematical that they cannot acquire these materials in the manner to which they are used with other raw materials such as steel and aluminum, by namely simply placing an order to buy on a commodity exchange. Potential suppliers of nanomaterials lack supply capacity, the finance to acquire it, and visible independent standards. These problems are not new and are solved by commoditizing nanomaterials (Section 11.3).

A novel, but hitherto untried, method of financing would be to exploit the possibilities of the era of social networking: nanotechnology companies that are being financed according to source (ii) are networked together in order to hedge underwriting risk. The key problem thereby overcome is that a fledgling nanotechnology company is usually based on a single technology, which makes it a high-risk investment because the technology might suddenly become obsolete. The networked collection of companies (that would, sensibly, share a common market for their products) then forms a basket of shares making up a "virtual" stock, which is what would actually be sold to raise funds.

11.2 Government funding

Possibly because of early British efforts (the UK National Initiative in Nanotechnology 1988–1993) and the massive US National Nanotechnology Initiative (launched in 2000), nanotechnology appears to have become indissociably linked with government funding, and today there is huge current investment in nanotechnology from the public domain. There are, however, interesting national differences (Table 11.1). A bald comparison of the absolute values is less revealing than key ratios: funding per capita (F/N) is indicative of the general level of public interest in pursuing the new technology, and the fraction of GDP spent on nanotechnology (F/G) indicates the seriousness of the intention. Despite the weakness of not knowing exactly how F has been determined, several general conclusions are interesting. Japan is clearly the leader, both in interest and intention. The USA follows in interest, and then France and Germany. France's strong interest is in accord with its current image as a powerhouse of high technology (the output of which includes the Ariane spacecraft, the Airbus, and high-speed trains). Switzerland's interest is surprisingly low, given its past lead as a high-technology country (but see Table 11.2). But when it comes to "putting one's money where one's mouth is," only Japan does creditably well. Switzerland in particular could easily afford to double or triple its expenditure [6]. And, impressive as the USA's contribution looks relative to that of Brazil, for example, it is barely half of the sum allocated to "funds for communities to buy and rehabilitate foreclosed and vacant properties" (taken as a somewhat random example of part of the US federal stimulus plan promulgated in February 2009).

The lower half of the table presents a less encouraging picture. The very low level of activity in Argentina, which formerly had a relatively strong science sector, is indicative of the success of the International Monetary Fund (IMF) in insisting on a substantial downscaling of that sector as part of its economic recovery prescription. In our modern high-technology era, this is simply *not* how to create the basis for a strong future economy, and reflects the archaic views that still dominate the IMF. Brazil's performance is also disappointing, given its aspirations to become one of the new forces in the world economy. Among these countries, only Malaysia reveals itself as a true Asian "tiger" able to take its place in the world nanotechnology community.

What of the effectiveness of the expenditure? If the main outcome of this kind of funding is papers published in academic journals, the ratio (P/F) of the number of papers P to funding is a measure of effectiveness (Table 11.2). By this measure the big spenders (Japan and the USA) appear to be less effective, and Switzerland's spending appears to be highly effective. This simple calculation of course takes no account of the existing infrastructure (the integral of past expenditure), nor to the extent to which funds result in products rather than papers. One nanotechnology paper costs about

Table 11.1 Government funding F (2004) for nanotechnology research and development, together with population N and GDP G.

Country[a]	$F^b/10^6$ €	$N^c/10^6$	$G^c/10^{12}$ €	F/N	F/G (%)
France	223.9	61	1.43	3.67	0.016
Germany	293.1	82	1.86	3.57	0.016
Italy	60.0	59	1.18	1.02	0.0051
Japan	750	128	3.03	5.86	0.025
Switzerland	18.5	7.4	0.25	2.50	0.0074
UK	133	60	1.49	2.22	0.0089
USA	1243.3	298	8.27	4.17	0.015
Argentina	0.4	38	0.12	0.010	0.00033
Brazil	5.8	188	0.59	0.031	0.0010
Malaysia	3.8	26	0.09	0.15	0.0042
Mexico	10	106	0.51	0.094	0.0020
South Africa	1.9	49	0.16	0.039	0.0012
Thailand	4.2	62	0.12	0.068	0.0035

[a] *The upper portion contains selected Category I countries (Section 14.3).*
[b] *Source: Unit G4 (Nanosciences and Nanotechnologies), Research Directorate General, European Commission.*
[c] *Source: Global Market Information Database. Euromonitor International (2008).*

Table 11.2 Number of papers (P, 2005) and P/F (Table 11.1).

Country	P^a	$(P/F)/10^{-6}$ €$^{-1}$
France	3994	18
Germany	5665	19
Italy	2297	38
Japan	7971	11
Switzerland	1009	55
UK	3335	25
USA	14750	12

[a] *R.N. Kostoff et al., The growth of nanotechnology literature. Nanotechnol. Perceptions 2 (2006) 229–247.*

5600 euros in France, which seems remarkably cheap, suggesting that expenditure on nanotechnology is underestimated (even if European Union funds are taken into account, it still amounts to less than 8000€).

In most countries, this public support [7] for research and development covers the entire range of the technology, with little regard for ultimate utility. What is lacking is a proper assessment of which sectors might best benefit from nanotechnology at its current level, frequently updated *pari passu* with scientific and technical developments. Such an assessment would make it possible to appraise the utility of current research, indicating into which sectors investment should be directed toward research, development, and innovation, and hence provide a better basis for public investment decisions, as well as being useful for private investors interested in backing nanotechnology-based industry. As it is, politicians are now having doubts about continuing the policy of lavish support for nanotechnology because the outcomes have been so meager. Although scientists and technologists continue to clamor for more government funding, it will surely be ineffective to merely increase spending on scientific and engineering research without rethinking some of the premises according to which it is carried out; in particular, a much more critical approach to its prosecution and the outcomes is needed.

The rationale for the public funding of scientific research is that it benefits all, hence it is a public good that cannot be monopolized—hence private investors will not pay for it and it will not be done if governments do not pay. In other words, the market price for the research outputs is too low to incentivize it. In the language of Adam Smith, self-interest cannot sustain universalities. This argument is closely connected with the Baconian "linear" model of wealth generation (Figure 2.1).

What is surprising at the beginning of the 21st century is that this view still persists so strongly after so many examples of its failure as well as empirical studies showing that the policy does not work. A well-known example of the superiority of private initiative is the birth of aviation. Samuel Pierpoint Langley benefited from generous federal funds but despite many years of effort he was well beaten by the Wright brothers, working on a shoestring with their own (private) resources. There are numerous other examples [8]. An extensive OECD study concluded that "...it is [business-performed R&D] that drives the positive association between total R&D intensity and output growth ... the negative results for public R&D ... suggest publicly performed R&D crowds out resources that could be alternatively used by the private sector" [9]. An excellent illustration of crowding out is provided as by the history of the Royal National Lifeboat Institution, a private charity founded in 1824 that maintains a network of lifeboats around the coastline of Britain. It is entirely financed by private donations and volunteer work and rescue operations are carried out free of charge. In the middle of the 19th century, however, it ran into financial difficulty and accepted an annual government subsidy but after a little more than a decade it was stopped, because the Institution found that private donations diminished and, moreover, the Institution suffered detriment because of the bureaucracy associated with receiving the government funds.

Yet further evidence comes from comparing the economic performance of countries, notably France and Germany, that generously supported scientific education and research starting in the 19th century, creating superb institutions that produced outstanding scientific work, with that of Great Britain, which did not [10]. Gross domestic product per capita even of Germany remained about 25% below that of Britain's from around 1800 until after World War II. In the USA, federal funding for science grew enormously (by almost three orders of magnitude), starting during

World War II, but barely affected the growth of the US economy (as measured by GDP per capita). In Japan, government investment in science has been marked by spectacular failures, such as the "Fifth generation" supercomputing project and the massive Key-TEC project that, when it was wound up, had achieved a return on investment of about 0.5%. The extraordinary technological sophistication of modern Japanese industry is almost entirely due to private research.

There is, indeed, good empirical evidence that corporate growth benefits from spending money on in-house research [11]. An interesting study of Pray et al. showed that (seed) manufacturers' return on R&D investment was 17%, a perfectly satisfactory figure. 94% of the value accruing from their research benefited the (very numerous) public (farmers, seedsman and, ultimately, all consumers of the grains), with only 6% being captured by the manufacturers (17 major private seed companies in India) [12]. This constitutes a model of how private company R&D benefits both the company and a large number of members of the public.

A major problem of government-funded research is that nowadays, in order to obtain it, one has to go through a long-winded, extremely bureaucratic application process involving a great deal of form-filling, the content of which must seem pointless to most scientists. Furthermore, it has become customary for the proposed research to have to be specified in great detail, to the extent that the major portion of the thinking part of the work has already been done by the time the application is ready to be submitted. If it is successful (i.e., the proposal is funded), the researcher merely needs to execute a detailed series of "workpackages," pass the specified "milestones," and produced prespecified results as "deliverables," enabling the funding agency to tick the appropriate boxes. The investigator receives a contract to carry out work that is the scientific equivalent of "painting by numbers"; in many cases it would surely suffice for a good laboratory technician to execute the "research" plan. Unfortunately nowadays it is typical for only about 10% of submitted proposals to be funded. Hence, 90% of the enormous preparation effort is wasted (and many government research agencies, such as Britain's Engineering and Physical Sciences Research Council, do not allow resubmissions). Furthermore, the whole procedure is extremely lengthy: from having the initial idea of the research to actually starting laboratory work (bearing in mind that the great uncertainty regarding the success of the applications means that recruitment of staff and procurement of equipment can only start once news is received that funding will be forthcoming) is typically two years. In any fast-moving field, this might mean that the proposed research has been overtaken by developments elsewhere before it even starts.

Yet another problem is that the decisions on what to fund are made by a committee. By definition, the majority of the members of any committee are working in the mainstream of the field and they will almost certainly vote for research proposals that also fall into the mainstream [13]. Therefore, government-funded research is always biased toward pedestrian investigations. It is a most unfortunate development that private philanthropy, which formerly used to be a very important source of finance for scientific research, has also become organized into formal bureaucracies that demand application procedures similar to those of the government research agencies [14].

Philanthropically funded research has likewise, therefore, evolved into a pedestrian activity. The reason for adopting the committee-based selection of research projects is the desire to avoid funding work that wastes precious public resources. As we have seen, however, the opposite is achieved. Furthermore, surely any real scientist would not wish to spend his or her time on a difficult problem that did not have the potential to change the world if a successful solution were found. A private charity may invite project proposals from the entire population and some weeding out is doubtless necessary, but most government research agencies impose strict eligibility criteria on applicants, who must generally be fully qualified scientists with established university positions—with, one would have thought, the professional ability to decide what research was worth pursuing.

It is worth pointing out that the scientist has far more in common with the mentality of the merchant than with that of the bureaucrat. This is well illustrated by a story of Thales of Miletus, generally considered to have been the world's first scientist [15]. Thales predicted from astronomical observations in the winter that next year's olive crop would be good, so without delay he raised some capital and bought the rights to all the olive presses. After the harvest he rented out the presses at a good profit. In other words: he made an observation (the positions of certain astronomical bodies); he made a hypothesis (that the positions he observed were correlated with conditions that would lead to an excellent olive harvest); he tested the hypothesis (buying the rights to the presses); and he measured the outcome (according to the profit he made). From this we can conclude that commercial practices are by no means inimical to science.

11.3 Endogenous funding

If both private venture capital and government funding are to be deprecated, what options remain? The answer is to be found in the very well-established exchange mechanisms for enabling trade to take place.

Exchanges such as the Chicago Mercantile Exchange and the London Metal Exchange and, in certain respects, stock exchanges around the world all have in common the feature that they allow standardized items to be bought and sold in a standardized fashion. Transaction costs (including those associated with price discovery) are thereby greatly reduced compared with the alternative mode of doing business, namely through bilateral contracts between supplier and buyer. An exchange therefore constitutes the most perfect practical form of a market. A simple fruit and vegetable market is a rudimentary kind of exchange, with the same qualities of transparency and openness that make the exchange an attractive medium for doing business.

"Standardized items" means that the items (e.g., particular grades of copper or wheat) fulfil published specifications. "Standardized item" is synonymous with "commodity" although a distinction is drawn between a commodity that is "commoditized"—capable of supply from several producers—and one that is produced by a sole or very few sources. An exchange will generally have the means for testing to ensure that items offered for sale on the exchange fulfil the specifications according to which they are offered. In order to ensure smooth running, both sellers and buyers have to register with (i.e., become members of) the exchange. By doing so they agree to abide by its rules (such as the prohibition of "insider trading," "open interest" disclosures to restrict attempts to corner a market, or restrictions on the size of intraday price fluctuations), which are strict, in order to ensure that both seller and buyer get the best possible deal.

One of the most important commercial innovations ushered in by exchanges was the concept of *forward selling*, in which the supplier is contractually bound to deliver a certain quantity of the commodity at a certain epoch in the future and the buyer is contractually bound to pay for it. The more sophisticated the technology behind the good the more important this is, because the preparation of the commodity demands ever more time and investment. Forward selling overcomes the often ruinous risk of production in advance of demand. A fisherman, for example, spends all night at sea but he does not know how many fish he will catch, nor does he know, when he returns to land, how many he will sell.

The exchange provides as nearly perfect a mechanism for adjusting supply to demand as is practically possible. If a good X (e.g., a certain grade of copper) is in short supply relative to the demand, this will be noticed by the suppliers and they will increase the price. The higher prices will attract more suppliers (e.g., those with more expensive means of production who would have been unable to sell at the previously lower price). It will also encourage forward selling, which provides the financial guarantee enabling investment to expand production facilities. (Nowadays, metal is nearly always sold when it is still in the ground as ore; as soon as the sale is agreed the miners rush to dig it out and put it through the extraction and refining processes.) Conversely, if there is a glut the price will fall and suppliers will withdraw until a balance is again achieved.

The alternative mode of business is well illustrated by the chemical industry. There is no exchange, even for the chemicals made in the largest volumes, and the market is typically characterized by enormous price differences between suppliers, excessive temporal price fluctuations, and extreme fluctuations of supply. Business has arranged itself to accommodate this endemic uncertainty, but a huge amount of effort is essentially wasted in the process compared with organizing an exchange, exacerbated by the inertia due to the large capital investment needed for many chemical production facilities. Presumably exchanges have never been organized in the chemical industry because suppliers believe they can command premium prices through the lack of transparency. The semiconductor industry has also traditionally eschewed exchanges for "chips" (very large-scale integrated circuits), perhaps because they were considered to be too sophisticated and special to be labeled "mere" commodities [16]. This viewpoint is, however, based on a fundamental misunderstanding. Just because a good fulfils published specifications and can be traded on an exchange does not preclude it from being sophisticated. (Indeed food products, whether commoditized or not, are, in terms of their internal structure, incredibly sophisticated—so much so that it is still impossible for humans to mimic them artificially.) In fact, "chips" are produced to strict specifications and in effect we have seen the commoditization of a range of microprocessors (e.g., the 386), without which their ubiquitous introduction into the domestic appliances, for example, would scarcely have been possible.

The present era of ultrahigh technology provides an interesting challenge to an exchange. It is easy enough to test a batch of gold, or copper to check whether it fulfils its specifications, and similarly with wheat (the testing of which does not, of course, involve detailed structural investigation at the molecular level). But the more sophisticated the product, the more difficult it is to specify it and test it for fulfilment.

The main commercial difficulty of nanotechnology at present is that there is a multitude of very small companies (many of them are university spin-outs), each making a different product in small quantities, which makes it extremely difficult for a potential user with a large-scale application (e.g., in the automotive industry) to do business. Take, as an example, carbon nanotubes as an additive for creating conductive polymers. The polymer manufacturer would need large quantities of a uniform specification with regular deliveries guaranteed. At present, no manufacturer is able to provide this. If, however, all the small suppliers joined an exchange and produced their nanotubes according to the exchange's specification, the polymer manufacturer might find his demands are met. Furthermore, through forward buying underwritten by the exchange process, the small suppliers would gain the financial guarantees enabling them to invest to expand their production facilities as needed. This form of financing provides a compelling incentive to the global investment community to *support the transaction* while ensuring a non-diluting (of ownership) source of capital for the small producer. As well as the direct benefits to both suppliers and buyers from facilitating individual trades, where the exchange could list many specifications, its operation would also lead to a general increase in the vitality of the industry, resulting in further growth, and so forth.

In the absence of an exchange, nanotechnology is likely either to remain an essentially academic activity with little commercial significance (excluding materials, such as carbon black, which were traded in large volumes long before the emergence of nanotechnology), or to follow the route adopted by the chemical industry (indeed, many large chemical firms are now actively pursuing nanomaterials, developing them both through their own research and through buying up small innovative companies). In the latter case, the nano-industry will be characterized by the same problems of price and supply fluctuations experienced by the chemical industry. But in the case of nanotechnology, because its products are more sophisticated than chemicals, as nanomaterials become more functional and smarter the difference between nanotechnology and the chemical industry will become more marked and the commercial difficulties of coping with the fluctuations might simply become so great that the industry never gains commercial viability.

Exchange-based trading also solves the problem of arranging insurance cover for nanomaterials and their shipment, since they are produced to published specifications and must satisfy relevant safety, health, and environment (SHE) provisions before they can be traded. Furthermore, every trade is tracked by the exchange and, hence, traceable.

Finally, beyond all such considerations, the exchange clearly reflects a democratic ideal for the equitable organization of human society, in which transparency, openness, and trust is a vital element to ensure universal participation in society. As technology becomes more and more sophisticated and widely diffused, ensuring that all members of society participate and feel that they have a stake in its continuing development appears to be essential to avoid a descent into anarchy.

11.4 Geographical differences between nanotechnology funding

Despite globalization, the fiscal environment is still a distinctively national characteristic. The three major poles of economic activity (the EU, Japan, and the USA) are quite sharply distinguished regarding expenditure on nanotechnology (research and technical development):

- Category I (Japan): roughly two-thirds private, one-third public.
- Category II (USA): roughly one-half private, one-half public.
- Category III (EU): roughly one-third private, two-thirds public.

Although the total expenditure in each of these three poles is roughly the same (around 4×10^9 CHF; again, the validity of this statement depends on what is included under "nanotechnology"), its effectiveness differs sharply. There can be no doubt that the Japanese model is the most successful. Solidly successful companies (without any magic immunity from the vagaries of the market) with immense internal resources of expertise have impressive track records in sustainable innovation according to the alternative model (Figure 2.2), but are well placed to develop nanotechnology according to the new model (Figure 2.3). Category II has several successful features, not least the highly effective Small Business Innovative Research (SBIR) grant scheme for funding innovative starting companies, and benefits from enormous military expenditure on research, much of which is channeled into universities. Category III is decidedly weak. There is an overall problem in that the fraction of GDP devoted to research and development in the EU is less than half that found in Japan or the USA. Moreover, what is spent is not well used. Many companies have been running down their own formerly impressive research facilities for decades (the clearest evidence for this is the paucity of top-ranking scientific papers nowadays emerging from European companies). Government policy has tended to encourage these companies to collaborate with universities, enabling them to reduce the level of public funding. Within Europe, there are immense differences between countries, however. Among the leading countries (Britain, France, Germany) France is in the weakest position. Traditionally anyway weak in the applied sciences, without a strong tradition of university research, and with its admirable network of state research institutes (the CNRS) in the process of being dismantled, there is little ground for optimism. In the UK, the level of innovation had become so poor that the government has virtually forced the universities to become commercial organizations, patenting inventions and hawking licenses to companies, and insisting on commercial outcomes from projects funded by the state research councils. Although university research is ostensibly much cheaper than company research, most companies seem nevertheless to have unrealistic expectations of how much they can expect to get from a given expenditure, in which they are anyway ungenerous to a fault. The British government avows the linear model (Figure 2.1), and is fond of emphasizing the importance of the research "base" as the foundation on which industrial innovation rests, but paradoxically is extremely mean about paying for this base, whose funds are cut at the slightest excuse, hence one cannot be optimistic about the future (although see Section 11.2 for an alternative view). In Germany, there is a strong *Mittelstand* of medium-sized engineering firms with many of the characteristics of Japanese companies. Furthermore, the state Fraunhofer institutes of applied sciences, along with the Max-Planck institutes (the equivalent of the French CNRS), are flourishing centers of real competence. If the EU were only Germany, one could be optimistic. The European Commission (the central administrative service of the European Union) seems to be aware of the problems, and has initiated a large supranational program of research and technical development, but the outcome is remarkably meager relative to the money and effort put into it. The main instrument is the "Framework" research and technical development program, but this is rather bureaucratic, easily influenced by dubious lobbying practices, and hence generally unpopular [17]. The bureaucracy is manifested by excessive controls and reporting requirements, brought in as a result of the generally deplorably high level of fraud in the overall EU budget (including agriculture, regional funds, etc.), which dwarfs the scientific activity *per se*, but all expenditure is subject to the same rules. There is little wonder that it has been concluded long ago that the "Framework" program actually hinders innovation in European industry [5], with no real evidence for improvement.

Given the outstanding success record of the SBIR grant scheme in the USA, it is astonishing that other countries have not sought to adopt it (Japan has its own very successful mechanisms, but unlike the situation in the USA and Europe as a whole, they are geared toward a far more socially homogeneous environment, as foreigners working in Japan cannot fail to notice). The situation within the European Union is especially depressing, marked as it is by ponderous, highly bureaucratic mechanisms and an overall level of funding running at about one-third of the equivalent in the USA or Japan. Switzerland manages to do better, but could actually easily exceed the (per-capita) effort of the USA and Japan (given that it has the highest per-capita income in the world). It is particularly regrettable that it has failed to maintain its erstwhile lead as a high-technology exporting nation, choosing instead to squander hundreds of milliards of francs on dubious international investments that have now (i.e., in 2008 and 2009) been revealed as worthless—or, more recently (since 2011), keeping in the currency artificially low. One can only wonder what might have been achieved had these same monies been spent instead on building up world-leading nanotechnology research and development facilities.

Rising neo-mercantilism, manifested by a combination of large state subsidies (including low-cost loans—often granted with little concern for near-term return on investment or overcapacity), national standards, preferential government procurement for national firms (a significant number of which are state-owned), and imposed requirements for technology transfer, is used by some countries, notably China and South Korea, to drive the growth of national innovation. The governments of these countries also encourage national enterprises to compete globally in strategic emerging industries with the help of loans and other support. Is this trend damaging to countries that practise and open trade policy? Past experiences of government-sponsored megaprojects (e.g., within the Japanese steel industry) have often turned out to be wasteful and weakening. Should it encourage a longer-term view in the open economies? It would seem regrettable if it triggered the general growth of neo-mercantilism all round. The open policy should be robust enough to withstand such threats, and history suggests that it can, although clearly a matter needs much more discussion than can be fitted in here.

References and notes

[1] de Haan S. NEMS—emerging products and applications of nano-electromechanical systems. Nanotechnol Perceptions 2006;2:267–75.
[2] Maine E, Garnsey E. Commercializing generic technology. Res Policy 2006;35:375–93.
[3] Maine and Garnsey, *loc.cit.*
[4] Parkinson CN, Parkinson's Law, p. 86 ff. Harmondsworth: Penguin Books; 1965.
[5] House of Lords Select Committee on the European Communities. Session 1993–94, 12th Report, *Financial Control and Fraud in the Community* (HL paper 75). London: HMSO; 1994.
[6] Incidentally, EU member states and other countries associated with their research and development program receive an additional 40% of the stated F.
[7] Not, interestingly, matched by public knowledge of nanotechnology, which remains remarkably scanty in most European countries.
[8] Including the hydrogen generation facility being built at Falkenhagen (Section 7.2.2), which is entirely funded by E.ON.
[9] *The Sources of Economic Growth in OECD Countries*. Paris (2003).
[10] British Prime Minister Robert Peel's dictum was "Of all the vulgar art of government, that of solving every difficulty which might arise by thrusting the hand into the public purse is the most illusory and contemptible."
[11] Mansfield E. Basic research and productivity increase in manufacturing. Am Econ Rev 1980;70:863–73; Grillich Z. Productivity, R&D and basic research at firm level in the 1970s. *ibid.* 1986;76:141–54.
[12] Pray CE et al. Private research and public benefit: The private seed industry for sorghum and pearl millet in India. Research Policy 1991;20:315–24.
[13] Gillies D. Lessons from the history and philosophy of science for research assessment systems. J Biol Phys Chem 2009;9:158–64.
[14] Ramsden JJ. Philanthropic support for science. J Biol Phys Chem 2012;12:87–8.
[15] The story is recounted in Aristotle's Politics. See T. Kealey, *loc. cit.*, pp. 88–9.
[16] A commodity is considered to be a good supplied without qualitative differentiation between suppliers across a market.

[17] A highly critical declaration launched in February, 2010 in Vienna entitled "Trust Researchers" attracted more than 13,000 signatures. It was addressed to the European Council of Ministers and the European Parliament.

Further reading

[1] Kealey T. Sex, science, and profits. London: Heinemann; 2008.
[2] McGovern C. Commoditization of nanomaterials. Nanotechnol Perceptions 2010;6: 155–78.

CHAPTER

Regulation

12

The regulation of industry began in the late part of the Industrial Revolution with the creation of agencies like the Railways Department of the United Kingdom Board of Trade (later called Her Majesty's Railway Inspectorate) following the Railways Regulation Act of 1840. One of the foremost railway engineers of the time, Isambard Kingdom Brunel, strongly opposed this development on the fully rational grounds that railway engineers "understood very well how to look after the public safety, and putting a person over them must shackle them. They had not only more ability to find out what was necessary than any inspecting officer could have, but they had a greater desire to do it" [1]. Few others had the high principles of Brunel, however, and the inspectors were duly appointed.

The concept of regulation as an engineering device was very much in the air at the time. James Watt had introduced a regulator, or governor, for his steam engines to enable them to run at a constant speed [2]. The Alkali Act of 1863 created a formal regulatory agency, the Alkali Inspectorate (the first Alkali Inspector was Robert Angus Smith, the Scottish chemist who pioneered the study of air pollution and acid rain). The creation was reactive: the infant heavy chemical industry in north-west England, notably the Leblanc process for making sodium carbonate (soda) from sodium chloride (salt) and calcium carbonate (chalk), released vast quantities of the highly corrosive hydrochloric acid into the atmosphere. Smith, believing that the best way to secure progressive improvements was to work with the industry rather than against it, established the consensual character of the Inspectorate, favoring compliance rather than enforcement, where negotiation and discussion predominated over formal, legal remedies [3].

Although the Alkali Inspectorate was highly successful—manufacturers complied with the need to eliminate emissions and found that the recovered hydrochloric acid was a highly profitable by-product—there is still controversy over whether regulatory agencies promote or depress the business they regulate. Whereas in the past industries have been well established by the time regulators have been appointed, in the intervening years the general public has become highly sensitized to the dangers of unprincipled and unregulated industry [4], and nanotechnology is now confronted by a barrage of regulation before the industry has really got under way.

In the past new technological advances, such as new substances that had some obvious beneficial use, were often adopted enthusiastically almost immediately after

Applied Nanotechnology, Second Edition. http://dx.doi.org/10.1016/B978-1-4557-3189-3.00012-9

their discovery with very little consideration given to adverse effects on human health or the environment. As a clear result of the accumulation of examples of adverse effects, a much more cautious approach has begun to be adopted, with the safety of new substances having to be demonstrated before they are allowed to be traded.

Such caution has long been an accepted part of pharmaceutical medicine (regulation represents, indeed, a form of risk management). Indeed, it might be considered an obvious corollary of the Hippocratic principle "primum non nocere", although the US Federal Food, Drug, and Cosmetic Act, which required the new medicines be tested for toxicity before being put on the market, was only passed in 1938. Nowadays extremely stringent procedures are in force to ensure the safety of medicines (at least for the majority—taking due account of genetic variation in drug metabolism is probably the next significant development in pharmaceutical safety), which apply equally to nanomedicine.

Apart from medicine, consideration of the safety of nanotechnology seems to have been placed under the umbrella of the chemical industry, perhaps because nanomaterials nowadays constitute the greatest volume of nanoproducts (ignoring nanoscale very large integrated circuits in computers), and many nanomaterials, especially nanoparticles, have traditionally been produced by chemists.

In the USA, chemicals are regulated by the Toxic Substances Control Act (TSCA) and the Occupational Health and Safety Act (OSHA). TSCA requires chemical manufacturers (and importers) to submit a pre-manufacture notification and risk assessment information to the Environmental Protection Agency (EPA). In principle these Acts could suffice to ensure the safe commercialization of nanomaterials. Nevertheless, conventional chemistry does not distinguish between a regular chemical substance and that same substance in the form of nano-objects. For example, the same material safety data sheet (MSDS) for graphite (carbon) could be used for fullerenes, carbon nanotubes, and graphene. Graphite is a rather harmless substance, whereas carbon nanotubes may be highly toxic, especially if inhaled. To deal with this potential problem, the EPA has a significant new uses rule (SNUR), meaning that even if a nanomaterial is chemically the same as an already registered bulk substance [5], it will need a fresh declaration.

By far the most draconian régime is being promulgated in the European Union, despite the fact that nanotechnology has been designated by the European Commission as one of the key enabling technologies (KETs) considered to be the drivers for future economic development. The basic regulatory instrument is the Registration, Evaluation and Authorization of Chemicals (REACH) system. This was only proposed in 2003 and has still not yet been fully implemented. REACH transfers the burden of proof that a chemical is safe from public regulatory agencies to the industry producing it. The European Union also has a new Classification, Labelling and Packaging (CLP) regulation. REACH will incorporate the precautionary (or "White Queen") principle, which is a firm part of official European Union policy [6]. The principle is that if something is *suspected* of bearing a risk of causing harm to the public or to the environment, in the absence of clear scientific evidence that it *is* harmful, then the burden of proof that it is *not* harmful falls on those wishing to introduce that something. There are no official guidelines concerning the degree of suspicion that would trigger the application of the precautionary principle, hence any legislation based on it is essentially weak and potentially the object of endless litigation. Perhaps, however, in practice its application will turn out to be workably consensual.

The main problem with the regulatory régime in the European Union is the plethora of regulations, and directives, recommendations, etc., potentially impinging on nanomaterials [7]. A small nanomaterials company will find merely keeping up with this enormous volume of paperwork an intolerable burden, yet failure to comply with some regulation possibly buried in one of the vast number of documents may lead to crippling litigation [8]. It is also unhelpful that some countries have introduced their own special regulations. The most notable is France, which has introduced the mandatory declaration of nanomaterials from 1 January 2013 [9]. This declaration is unlikely to be effective in tracing nanomaterials because it only obliges anyone producing more than 100 g of a nanomaterial to file an annual return. A further problem is that the European Union has promulgated its own definition of a nanomaterial, which is different from that reached by ISO after 3 years of laborious consensus-seeking. Finally, there are grave doubts concerning the quality of the scientific advice provided to the European Commission. In common with most of the internal workings of the European Union the procedure whereby people (in this case, scientists) are appointed to advisory committees, such as the Scientific Committee on Emerging and Newly Identified Health Risks (SCENIHR), is opaque. Occasionally, independent external experts are invited to comment on the reports of these committees and as a rule the comments are highly critical. It is not untypical for the Commission officials to agree with the criticism—but it is not considered to be appropriate for them to directly criticize the work of experts that have been officially appointed! As a net result, small companies (which are supposed to constitute the driving force for nano-innovation) are bogged down; only the largest companies have the resources to lobby the European Commission to ensure that their commercial interests are not damaged. Under this régime, any emergence of a competitive industry within the member states of the Union will be in spite of, not because of it; it is regrettable in the extreme that so much effort and energy has to be expended on artificially erected obstacles of this nature.

Globally international activity is generally confined to guidelines and recommendations. Examples are the World Health Organization's "Guidelines on Nanomaterials and Workers Health", the documents produced by the OECD's Working Party on Manufactured Nanomaterials, and the United Nations' Globally Harmonized System of Classification and Labelling of Chemicals.

Referring back to Brunel's criticism of regulation, by far the most effective solution to ensuring the safe worldwide deployment of nanomaterials would be to let the industry regulate itself. This can very easily be accomplished through the medium of a commercial exchange (Section 11.3). Due diligence is applied to all companies wishing to become members of an exchange, ensuring that only reputable firms are allowed to do so. All trade is in materials that comply with published standard specifications, and every trade benefits from downstream audit sequencing (DAS), also known as "track-and-trace": each individual trade has a unique number that is sequenced from producer to consumer and is entirely compatible with CLP. Ensuring that all nanomaterials are exchange-traded makes the new French reporting system completely redundant, and is far more effective since every trade is essentially logged when it occurs, rather than consolidated into an annual return.

Unlike conventional materials, it is often extremely difficult to trace nano-objects once they are incorporated into a product, let alone once they are released into the environment. For example, silver nanoparticles are now incorporated in several textile products, such as wound dressings and socks, as a bactericide. If the textile is laundered many of the particles escape into the waste water and could end up widely dispersed in the environment. The difficulty of tracing nano-objects is above all due to their minute size and their fugitive nature (which is a consequence of their minute size). Regulations may forbid the release of silver nanoparticles into the environment (because of the risk of destroying bacteria essential for a healthy ecosystem), but how is such a regulation to be enforced if the nanoparticles cannot be detected? Silver would at least be an unusual element to find in most soils, but iron nanoparticles, which are also finding an increasing number of applications, are likely to closely resemble natural components of many soils. Possibly in the future the discipline of "nanoforensics" will be developed [10].

Risk is the product of hazard and exposure. Certain nano-objects may be very hazardous if they are inhaled (and suitable precautions must be taken during their manufacture). But, once they are incorporated into another product, the probability of exposure may become negligible. An example is the use of carbon nanotubes in composite materials. The nanotubes are dispersed in a polymer matrix. For most structural applications, the nanotubes will never be released and, hence, the overall risk is negligible. Nevertheless, the composite is a nanomaterial according to the standard vocabulary developed by ISO (and according to the official definition of the European Union) and may, therefore, be subject to very restrictive conditions of use according to present or proposed regulations.

Although in some countries (such as the UK) the government is required to carry out assessments (called impact assessments, IA) of the likely costs, benefits, and impacts of any legislation it implements [11], little assessment appears to have been done at EU level, which is the main locus for the regulation of nanotechnology within Europe. The present situation appears to be rather fluid, and stakeholders would be well advised to energetically investigate exactly what is happening that potentially impacts on their activities, and allocate resources for defending their interests.

References and notes

[1] Quoted in Rolt LTC. Isambard Kingdom Brunel. London: Longmans, Green & Co.; 1957. p. 217. Seven years later (1848), Brunel was obliged to express similar sentiments with respect to the Royal Commission on the Application of Iron to Railway Structures: "...it is to be presumed that they will lay down, or at least suggest, 'rules' and 'conditions' to be observed in the construction of bridges, or, in other words, embarrass and shackle the

progress of improvement tomorrow by recording and registering as law the prejudices or errors of today. No man, however bold or however high he may stand in his profession, can resist the benumbing effect of rules laid down by authority. Devoted as I am to my profession, I see with fear and regret this tendency to legislate and rule."

[2] Maxwell JC. On governors. Proc R Soc 1867–1868;16:270–83.

[3] This tradition persisted until the 1970s, when the Alkali Inspectorate was absorbed into the Health and Safety Executive (1975) and lost its independence.

[4] See Michaels. Doubt is their product. Oxford: University Press; 2008, for many case studies.

[5] This begs the question of the meaning of "chemically the same". In terms of elemental constitution, graphite and carbon nanotubes are indeed the same. Most of chemistry is, however, concerned with compounds rather than elements, in which the bonding plays a primordial rôle (and even the bonding between elements very often determines their nature: the difference between diamond and graphite is in the bonding between the carbon atoms). Most or all of the atoms in a nano-object are on the surface of the object, hence their bonding is different from that of their bulk congeners.

[6] Article 191 of the Treaty on the Functioning of the European Union. See also the Communication from the Commission on the Precautionary Principle (COM(2000)1, February 2000). Neither of these documents, however, actually defines the precautionary principle. Perhaps it is considered to be self-evident.

[7] See, for example, the list of references in the ObservatoryNANO "Development in Nanotechnologies Regulation & Standards" Report No. 4 (April 2012).

[8] A key document appears to be the "Communication from the Commission to the European Parliament, the Council and the European Economic and Social Committee: Second Regulatory Review on Nanomaterials" (COM(2012)572, October 2012).

[9] Decree No. 2012-232 (17 February 2012) defining the modalities of application, with the information to be declared specified by the decree dated 6 August 2012.

[10] At present this term seems to be exclusively used to mean the application of nanotechnology in forensic science. Such applications might include the use of scanning probe microscopes to investigate materials found at the scene of a crime, and biologically functionalized nanodots as markers for certain DNA sequences.

[11] In the UK, IA are carried out by the Better Regulation Executive (BRE), part of the Department for Business, Innovation & Skills.

Further reading

[1] Feitshans IL. Forecasting nano law: defining nano. Nanotechnol Perceptions 2012;8:17–34.

CHAPTER

Some Successful and Unsuccessful Nanotechnology Companies 13

CHAPTER OUTLINE HEAD

13.1 NanoMagnetics . . . 161
13.2 MesoPhotonics . . . 162
13.3 Enact Pharma . . . 163
13.4 Oxonica . . . 163
13.5 NanoCo . . . 164
13.6 Hyperion . . . 164
13.7 CDT . . . 165
13.8 Q-Flo . . . 166
13.9 Owlstone . . . 166
13.10 Generic Business Models . . . 167

Commercial nanotechnology is epitomized by small companies, typically spun out of a university research department. That is not to say that large companies do not engage in nanotechnology—many of them do, but they are (of course) far smaller in number. This chapter will look exclusively at small companies, which constitute the main innovative force in the field. Large companies often get involved in nanotechnology by buying up a small company at a promising-looking juncture. An internal nanotechnology project within a large company might share many of the characteristics of a small independent company, but information about its history is usually a confidential, internal matter of its corporate owner, which is another reason for focusing on the small companies here.

Despite many commercial surveys proclaiming multibillion dollar markets for nanoproducts, the record of success of small companies is, on the whole, rather dismal. Faults seem to be repeated again and again. There would appear to be nothing more straightforward than exploiting a brilliant idea that is unique and technically superior to all existing materials for doing the same job. Time and time again, however, a material is launched without there being any prospective buyers. Since much nanotechnology is generic, a company might be formed to make, for example, nanoparticles. The company's unique know-how resides in its process for making the nanoparticles,

Applied Nanotechnology, Second Edition. http://dx.doi.org/10.1016/B978-1-4557-3189-3.00013-0

which it has demonstrated in the laboratory. Prospective buyers are then told "we can make any kind of nanoparticle that you like". If a customer does not come with the specification of what nanoparticle it needs (presumably to incorporate into another product), it is unlikely to find such a declaration helpful. Contrast this with the product range of the German company Freudenberg, a world leader in industrial lubricants and in elastomers: in each of these areas, Freudenberg sells about 30,000 different kinds of products. Each one corresponds to a well-defined demand. The nanoparticle company might easily be able to make 30,000 different kinds of nanoparticles, but does not succeed in selling even one kind. Naturally it is not entirely fair to compare an established company that has attained its leading position after many years of sustained effort with a recent start-up. But the principle of manufacturing that for which there is demand nevertheless holds.

The second failure is choosing the wrong model for financing the company. This is covered in more detail in Chapter 11; in brief, the "neoclassical" model that is taught in business schools and has become very widely disseminated is to start with some seed capital, perhaps from the inventors themselves and a close circle of friends and relatives, then procure "angel" finance and finally move on to venture capital and/or private equity. At that stage the founders have essentially lost control. The priorities of the board now become very different. Making and selling the brilliant new product, which would ensure prosperity in the long term, is subordinated to extracting maximum short-term value from whatever the company has to offer. As a result, most nanomaterials companies seem to adopt the business model of simply licensing technology, managing projects and, possibly, further research and development. The actual manufacturing is outsourced. The true classical model is what was developed and honed in the early years of the Industrial Revolution—the community of producers and consumers organized an exchange for selling and buying, whereby forward selling provided the capital to finance scaleup of production. This exchange system has its roots in agriculture—an arable farmer necessarily has to buy seed in advance of the harvest, and forward selling the crop is a straightforward and efficient way of enabling this to be done (cf. Section 11.3).

Some of the other warning signs of company ill health are: an excess of executive directors and managers over employees, which seems remarkably often to account for the high cash burn rate of many start-ups; too much focus on the technology itself without considering the user interface (this is more of a problem with nanodevices rather than nanomaterials); and a simple lack of entrepreneurial vision about why the company exists—what could be called the company "religion"—and where it is going and how to get there. Getting this conceptual basis right is probably essential to ensure that all staff, whether directors or employees, have the determination and capacity for hard work to get the company off the ground.

In the first five sections of this chapter the histories of the (initially British) companies highlighted as model examples of entrepreneurial nanotechnology in the Taylor Report [1] are briefly recounted. Only two are still in existence and only one of them can, at the time of writing, be said to be successful. The remainder of the chapter updates information about the four companies described in the first edition of this book: two were medium sized and two were small, and in each size category one was very successful and one perhaps a haltingly successful.

Key ingredients of success are evidently:

- Focusing on a single application.
- Launching as downstream a product as possible.
- Making a prototype to demonstrate value.
- Having dedicated staff.

Spin-out companies are often tempted to economize by continuing to use university facilities and part-time staff, but this seems to ensure that the necessary pressures to succeed never surmount what might be a critical threshold. Doubtless location in a thriving center of high technology is important (but might be becoming less so in the age of the internet). Given the novelty of the upstream product, persuading downstream companies to incorporate it into their final product, with all the attendant expense of redesign (even if the upstream product is merely substitutional) may be even more expensive than pursuing the downstream product in-house. Here, a rational basis for estimating the costs is important (although difficult). And, even if the downstream client is a partner, it may still be difficult to obtain accurate information about key attributes. Finally, as already mentioned, it is known that the further upstream one is positioned, the harder it is to capture value from any specific application, which diminishes the attraction for investors.

13.1 NanoMagnetics

This company's product was nanoscale magnetic particles (DataInk™) made very uniformly by deriving them from the natural iron-storage protein ferritin. It was claimed that by coating magnetic media with these novel particles, their storage density could be raised by two orders of magnitude. "NanoMagnetics started in the research laboratories at Bristol University. Over the past couple of years the company has filed several patents, raised £6.7 million, and recruited a high-powered CEO. Earlier this year the UK Minister for Science and Innovation cut the ribbon at its new purpose-built 10,000 sq. ft laboratories in Bristol [1]."

The claim sounds plausible but currently the feature limiting the usable storage density is the size of the reading device, which will not change even if the magnetic particles are superior. Furthermore, a very similar research project has been running for some years at the Nara Institute of Science and Technology (NAIST), sponsored by Panasonic [2]. The project successfully demonstrated the material, but Panasonic decided not to exploit it due to some incompatibilities in the production process.

The company was considered to be a star in the UK firmament of entrepreneurial nanotechnology. It was founded in 1997 by Eric Mayes, then doing a PhD at the University of Bath, and Nick Tyler, a former investment banker, with very limited resources. In 1999 they won a Department of Trade and Industry (DTI) "Smart" award worth £133,000 and secured a further £650,000 in venture capital, which enabled them to set up a laboratory within the Department of Physics at Bristol University and establish a research team. The venture backing was later increased to $10 million. The Royal Society of Chemistry presented the cofounder with the Chemistry Entrepreneur of the Year Award in 2004. The company's business model was to carry out research and development, license its technology, and outsource all production. It had about a dozen full-time employees.

Administrators were appointed on 27 January 2006 and the company was formally dissolved on 27 January 2011.

Interestingly, another little and almost eponymous company, NanoMagnetics Instruments, was founded in 1998 to develop high-resolution scanning Hall probe microscopes for room temperature magnetic imaging applications. A low temperature microscope was launched in 2001 and a low temperature magnetic/atomic force microscope launched in 2005, followed by a room temperature version achieving sub-10 nm magnetic resolution in air. Another novel instrument, a noncontact atomic force microscope, capable of achieving atomic resolution on mica in water, was launched in 2010. This company has been quietly forging ahead, without government awards and without venture capitalists, providing specialist instruments to the nanoscience community, which evidently demands what they produce.

In the USA, NanoMagnetics Ltd. produces "Nanodots" in Wilmington, Delaware; it dubs them "Lego for the 21st century". This is a consumer product, based on centimeter-scale magnetic spheres, which can even be considered a child's toy, although it is also being used by artists for making elaborate sculptures.

13.2 MesoPhotonics

This company was launched in 2001 to design and develop photonic crystal nanodevices made using conventional semiconductor (silicon) processing technology. It was spun out of the Department of Electronics and Computer Science at the University of Southampton by Prof. Greg Parker, who retained his university post while becoming technical director of the company, and received an investment of £2.8 million from BTG. Proof-of-principle devices (building on almost a decade of prior research within the University) were envisaged by mid-2002 with demonstrator devices going out to potential customers in early 2003 [1]. £5.5 million was secured in a second round of venture financing in 2004. In January 2005 the company launched "Klarite", a nanostructured surface material exploiting surface-enhanced Raman scattering (SERS).

The company's business model was to undertake modeling, design, testing, applications, engineering, and sales while outsourcing semiconductor and device fabrication. It had about 20 full-time employees.

In 2007 the SERS substrate technology was sold to D3 Technologies (a collaborative venture between the University of Strathclyde and Renishaw) for £0.85 million. By 2013 MesoPhotonics appear to be inactive.

13.3 Enact pharma

Two existing companies merged in 2000 to form this development company focused on cancer and neurological diseases, which went on to raise over £5 million in equity investment and was listed on Ofex. Its technology portfolio included cancer therapy and specially treated biodegradable polymers able to provide a chemical pathway for nerve cells to grow along. The company was based on the Porton Down Science Park. By 2003 its drug "Voraxeze", to combat side-effects in cancer treatment, was in late stage development (with launch envisaged in 2005 or thereafter); the company had debts of over £1 million. In the same year it was bought by Protherics, a company specializing in making snakebite treatments.

13.4 Oxonica

Originally called Nanox, the company was spun out of Oxford University's Department of Materials in 1999 by Prof. Peter Dobson and Dr. Gareth Wakefield with a starting investment (private sources) of £750,000. Angel investors contributed £100,000 in 2000, the name was sold to another company for a similar sum and changed to Oxonica in 2001 and further angel finance amounting to £540,000 was secured. In that same year Dr. Kevin Matthews was appointed as a full-time CEO and found "a business with no real product focus, lots of research but no real commercial focus, and low morale". Venture capital funding of £4.2 million was secured in 2002 followed by a rights issue bringing in £4 million in 2004, in which year the company also won a DTI "Smart" award worth £450,000. A further $2.5 million was received in venture funding in 2005. It was floated on the Alternative Investment Market of the London Stock Exchange (AIM) between 2006 and 2009. The business model is technology and marketing, with production being outsourced. Its main offerings are "Envirox", a nanoparticulate catalyst for increasing the combustion efficiency (between 5 and 10%) in diesel engines, originating elsewhere but developed into a product by Oxonica and trialed in hundreds of buses in 2003–4; an inorganic nanoparticulate sunscreen ("Optisol") that does not form highly reactive free radicals upon illumination, based on an invention of Prof. Dobson (patent owned by Oxford University); a silver nanoparticle SERS product ("Nanoplex") from the University of Strathclyde, intended for biomarker detection; and a polymer stabilizer ("Solacor"), intended for use in cosmetics.

An "appallingly drafted" patent agreement with Neuftec resulted in Oxonica losing a High Court case (and a subsequent appeal) over Envirox, following which Oxonica's Energy division was sold in 2009. Oxonica Materials Inc. was sold to Cabot Corporation in 2010, generating $4 million. Given the uncertainties over the health hazard associated with the emission of large quantities of semiconductor nanoparticles in engine exhausts, this was probably a prudent divestment. The company's primary objective is now "to return surplus capital to shareholders" based on license agreements for the SERS and sunscreen products. It still proclaims itself as "leaders (*sic*) in nanotechnology" on its website.

13.5 NanoCo

NanoCo Technologies Ltd started with one full-time member of staff when it was spun out of his research group at Manchester University in 2001 by Prof. Paul O'Brien together with Dr. Nigel Pickett. The company's product is quantum dots for security applications. They were developed in collaboration with a major company who desired them for anti-counterfeiting purposes and which initially approached Prof. O'Brien. The special, innovative technology of the company is a molecular seeding process that enables the bespoke manufacture of nanodots on a commercial scale. The company is also at the cutting edge of chemical functionalization of the particles for biomedical and other purposes. The company was admitted to AIM in 2009. Since its inception it has raised £4.1 million of private equity funds; these investors are now the most significant shareholders. Its main research and manufacturing facility remains in Manchester; it also has a facility in Japan.

Although the company attracted some negative publicity in 2011 when it threatened to move its manufacturing overseas unless the UK government provided some support (no support was forthcoming and the company remained in Manchester), it has essentially got things right: based on demand coming to the company rather than launching a product for which no proven demand exists; stringently controlling costs (especially staff) in the early stages; and manufacturing as well as doing research and development. The only regrettable feature is that the company is now controlled by private equity.

13.6 Hyperion

Hyperion (unfortunately a rather popular name for businesses, presumably because it means "the high one", the name of one of the twelve Titans of Greek mythology) was founded in 1981 to develop carbon filament-based advanced materials for a variety of applications. Located in Cambridge, MA, it developed its own process for the fabrication of multiwalled carbon nanotubes (MWCTs), its key intermediate product. By 1989 it could make them in-house on a fairly large scale and to a high level of purity. The problem then was to choose a downstream application. In the absence of prototypes, it widely advertised its upstream product with the aim of attracting a partner. Its first was a company that had developed a competitive polymer automotive fuel line as a substitute for the existing steel-based technology, but still needed to make the polymer electrically conducting (to minimize the risk of static electricity accumulating and sparking, possibly triggering fuel ignition). Dispersing MWCTs in the polymer looked capable of achieving this, and by 1992 Hyperion had developed a process to disperse its material into the polymer resin, meanwhile also further upscaling its process to reach the tonne level. Related applications followed from the mid-1990s onward—conductive polymer automotive mirror casings and bumpers, which could be electrostatically painted along with the steel parts of the bodywork and, hence, fully integrated into existing assembly lines. The company moved slightly downstream by starting to compound its MWCTs with resin in-house. Efforts to diversify into structural aerospace parts did not succeed in demonstrating adequate enhanced value to enter the market, but the company did successfully break into internal components of consumer electronics devices. Research into supercapacitors and catalysts was pursued with the help of government funding. By the year 2009 Hyperion had filed over 100 patents. The product line remains based on carbon nanotubes dispersed in resin to make it conductive. In 2011 the firm concluded a patent license agreement with Bayer MaterialScience, enabling Bayer to use Hyperion's CNT within a defined field.

Today the company is called Hyperion Catalysis International and proclaims itself as the world leader in carbon nanotube development and commercialization. It is trying to enter the market for precious metal catalysts using a network of fused carbon nanotubes.

13.7 CDT

Cambridge Display Technology (CDT) was founded in 1992 as a spin-out from Cambridge University (UK), where during the preceding decade polymer transistors and light-emitting polymers had been invented (the key polymer electroluminescence patent was filed in 1989). CDT's objective was to manufacture products for flat-panel displays, including back lighting for liquid crystal displays. It soon became apparent that a small company could have little impact on its own, hence it abandoned in-house manufacturing and sought licensing arrangements with big players such as Philips and Hoechst (finalized by 1997), and in 1998 embarked on a joint venture with Seiko-Epson Corp. (SEC) to develop a video display. Other strategic allies included Bayer, Sumitomo, Hewlett-Packard, and Samsung. CDT continued patenting (end-products developed with allies were included in the portfolio), but R&D costs remained huge, far exceeding license revenues. In 2000 the company was acquired by two New York-based private equity funds. This caused some turbulence: the departure of the energetic CEO (since 1996) and the decision of the founder, Richard Friend, to form a new company on which he focused his continuing research efforts. CDT then decided to recommence manufacturing and released an organic light-emitting diode (OLED) shaver display in 2002, but an attempt to extend this to the far more significant cellular telephone market came to nought and the commercial-scale production line was closed in 2003, retaining only the ability to make prototypes. The company thus reverted to the licensing mode. By 2003 it held 140 patents, generating \$13–14 million per annum, compared with annual running costs of *ca* \$10 million (the company had 150 employees at that time). The strategy of getting the technology into small mobile displays in the short term, and aiming at the huge flat-panel market (estimated as $30 milliard annually) in the medium term has remained attractive to investors despite the ups and downs, and the company went public on the NASDAQ in 2004. In 2007 CDT was acquired by Sumitomo Chemical (which had started its research activity in polymer light-emitting diodes in 1989), of which it became a wholly owned subsidiary. Since then the company has been active in forging links with overseas partners—with Semprius in 2008, and in 2009 CDT joined the Center for Organic Photonics and Electronics at Georgia Tech (USA) as an industrial affiliate. In 2011 CDT signed a five-year intellectual property pipeline licensing agreement with the Organic Nano Device Laboratory of the National University of Singapore to facilitate the commercialization of research results.

13.8 Q-Flo

Q-Flo was founded in 2004, also as a spin-out from Cambridge University (UK), in order to commercialize a novel process for making carbon nanotube (CNT) fiber (at a cost potentially one-fifth that of then-current industrial CNT fiber) as a very strong material in the form of a textile or a film. Favorable electrical properties were reflected in envisaged applications in supercapacitors and batteries. Other opportunities include bulletproof body armor, shatter-proof concrete, ultrastrong rope, tires, and antennas. However, the company is too small to be able to afford to make prototypes for value-demonstration purposes, but in their absence cannot attract the investment needed to be able to afford to make them. The key resource-building cycle (see Figure 11.2) cannot therefore start turning. Because the company is so small, none of its seven employees work full time for Q-Flo, which also limits the intrinsic dynamism of the available human resources.

A promising development has been the joint venture company TorTech Nano Fiber Ltd formed in 2010 with Plasan Sasa Ltd to scale up Q-Flo's technology for industrial production (presently output is limited to 1 g/day), for body armour and other defense applications, for which Plasan has exclusive sales and marketing rights.

13.9 Owlstone

Owlstone was also founded in 2004 as a spin-out from Cambridge University (UK). Its technology is nanoscale manufacturing to produce a microelectromechanical system (MEMS) gas sensor, based on field-asymmetric ion mobility spectrometry (FAIMS). This generates a "fingerprint" for any gas or vapor entering the sensor, which is matched against a collection of standard fingerprints. The device is several orders of magnitude smaller than existing competitors and detection takes less than 1 second. The company's first investor was Advance Nanotech, which acquired a majority interest, but after other companies owned by Advance Nanotech failed to reach expectations, Owlstone took over its erstwhile owner.

The original aim was to make the FAIMS chip and sell it to sensor suppliers, leaving it to them to incorporate it into their products. However, the uniqueness of the device meant that outsourcing production of the chip alone would incur high development overheads with general foundries in any case, hence it was decided to aim instead to produce the finished downstream sensor. With the help of SBIR funds (fortunately for this purpose Advanced Nanotech was registered in both the UK and the USA) the first production-model sensor was launched in 2006. Further products were subsequently launched with partners already in the market. Revenue in 2008 was expected to exceed $2 million, and since then the company has continued to expand, entering the pharmaceutical (2009) and crude oil (2012) domains.

13.10 Generic business models

Wilkinson has identified four generic business models (Figure 13.1), all beginning at the most upstream end of the supply chain, but extending progressively downstream.

The typical temporal evolution of new technology is shown in Figure 13.2; it is based on a model of the printed circuit board industry, and so far has fitted the observed course of events in microsystems. It can reasonably be considered as a model for nanotechnology as far as its substitutional and incremental aspects are concerned. Insofar as it is universal and radical, however, prediction becomes very difficult.

If one examines in more detail the early stages, it appears that there might be a gap between early adopters of an innovation and development of the mainstream market

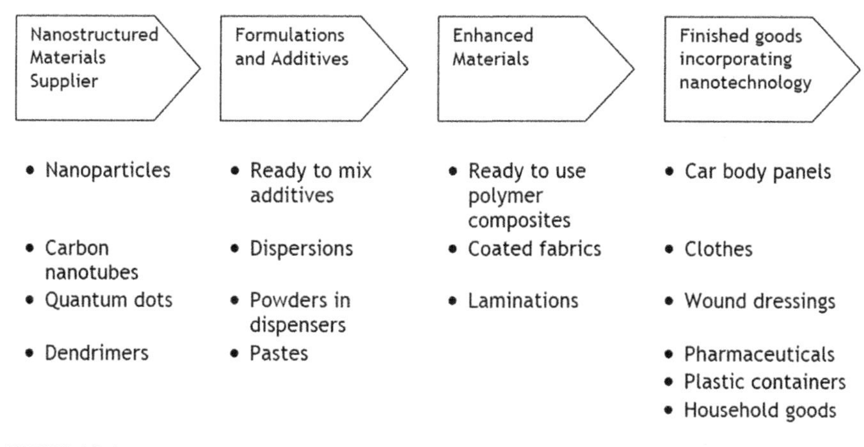

FIGURE 13.1

Generic business models for nanomaterial suppliers. Model A (e.g., Thomas Swan) produces only nanostructured materials. Model B (e.g., Zyvex) produces nanostructured materials and formulates additives. Model C (e.g., Nucryst) produces nanostructured materials, formulates additives, and supplies enhanced materials. Model D (e.g., Uniqema Paint) produces nanostructured materials, formulates additives, makes enhanced materials and finished goods incorporating those materials. Reproduced from J.M. Wilkinson, Nanotechnology: new technology but old business models? *Nanotechnol. Perceptions* 2 (2006) 277–281 with permission of Collegium Basilea.

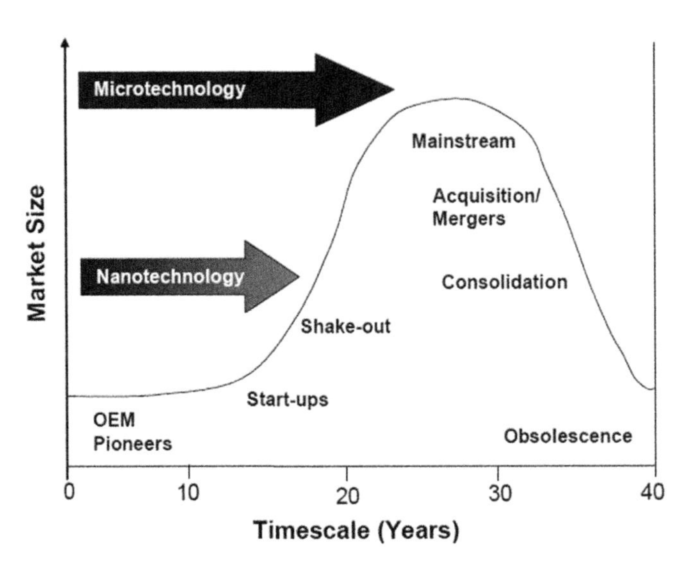

FIGURE 13.2

Generic model proposed for the temporal evolution of nanotechnology companies (originally developed by Prismark Associates, New York for the printed circuit board industry; it also seems to fit the evolution of the microtechnology industry). Reproduced from J.M. Wilkinson, Nanotechnology: new technology but old business models? *Nanotechnol. Perceptions* 2 (2006) 277–281 with permission of Collegium Basilea.

[3]. The very early market peaks and then declines, followed by the mainstream development as shown in Figure 13.2. This feature underlines the importance of patient investors in new high-technology companies.

Most nanotechnology start-ups use model A or B, sometimes C (the boundary between B and C is fuzzy): they must rely on finding other companies able to act as customers or partners in order to deliver goods to the marketplace (e.g., Nucryst has partnered with Smith & Nephew). Models C and especially D require large quantities of (venture) capital.

References and notes

[1] New Dimensions for Manufacturing: A UK Strategy for Nanotechnology (report of the Advisory Group on Nanotechnology Applications, chaired by John Taylor). London (2002).

[2] Igarashi M, et al. Directed fabrication of uniform and high density sub-10-nm etching mask using ferritin molecules on Si and GaAs surface for actual quantum-dot superlattice. Appl Phys Express 4 (2011) 015202; M. Uenuma et al., Resistive random access memory utilizing ferritin proteins with Pt nanoparticles. Nanotechnology 2011;22:215201.

[3] Moore GA. Crossing the chasm. New York: Harper Business; 1991.

CHAPTER

The Geography of Nanotechnology

14

CHAPTER OUTLINE HEAD

14.1 Locating Research Partners 171
14.2 Locating Supply Partners 173
14.3 Categories of Countries 173
14.3.1 Nanotechnology in the Developing World 174

Although we live in an era of globalization, and although science has always been global in spirit, there are nevertheless huge differences in the degree and history of technological development in different countries. This chapter explores some of these differences in activity—nanotechnology in a global manufacturing context.

A good place to start is by comparing the outputs of nanotechnology research papers of the leading countries (Figure 14.1). The comparison may be considered to be misleading because it takes no account of the different sizes of the countries. Figure 14.2 compares the output of research papers per capita. The ranking is very different. Most remarkable of all is the position of South Korea. The Korea Nanotechnology Initiative was only launched in 2001; according to the master plan for the initiative's second phase spanning the decade from 2006 to 2015, Korea aims to become one of the world's top three nations in global nanotechnology competitiveness by 2015. Competitiveness is not yet defined by a standardized procedure but, according to Figure 14.2, it was already in first place by 2009! The country is no less advanced industrially, mass producing memory devices with features at the 30–40 nm scale, and the website "nanowerk" lists 30 nanotechnology companies in South Korea. Nor is the initiative driven by empty platitudes, but by a visionary coherence that sees such aspects as the unity and generality of principles, nanotechnology as the ultimate technology for material and system production, and as the most efficient length scale for manufacturing as powerful motivations for attaining excellence in the field.

Remarkable in a different way is Germany, which has achieved almost the same rank (according to the publications) without any grand overarching national initiative (the main instrument is an "Innovation Initiative"), and the same for France

Applied Nanotechnology, Second Edition. http://dx.doi.org/10.1016/B978-1-4557-3189-3.00014-2

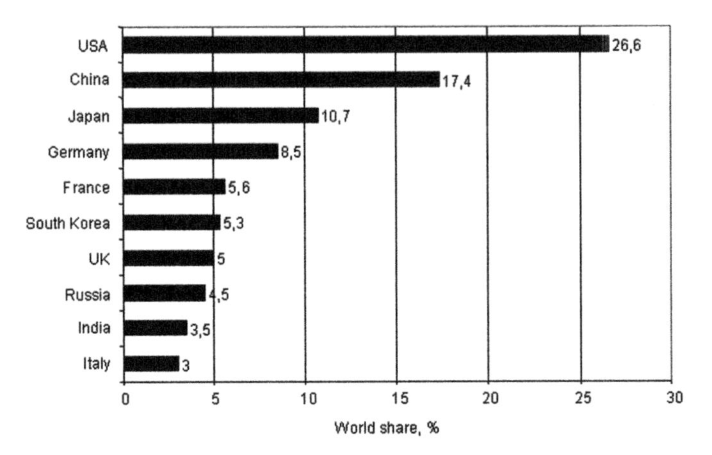

FIGURE 14.1

Different countries' proportions of publications dealing with nanotechnology (from 1990 to 2009).

Reproduced from A.I. Terekhov, Developing nano research in Russia: a bibliometric evaluation. Nanotechnol. Perceptions 7 (2011) 188–198 with permission of Collegium Basilea.

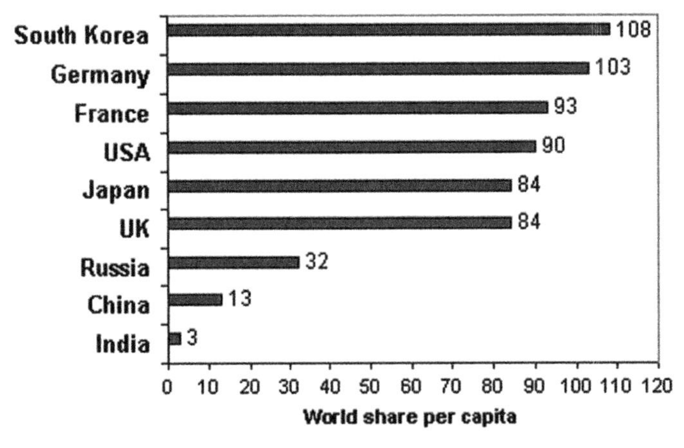

FIGURE 14.2

Different countries' proportions per capita of publications dealing with nanotechnology. The data from Figure 14.1 has been divided by the countries' populations in 2007.

From the Pocket World in Figures 2007, London: The Economist, except for South Korea which is not included in that compilation; its population in that year has been estimated from data provided by the National Statistical Office of the Republic of Korea.

too (although the Académie des Sciences recommended initiating a "vast program"). Perhaps two centuries of state investment in science and technology (cf. Section 11.2) are finally yielding dividends! Japan and Britain are slightly disappointing, given that they really initiated nanotechnology (Japan in 1974, see Figure 1.1 and Britain in 1988 with the first nanotechnology initiative, but it only lasted 5 years and nothing immediately followed it). Russia must rank as very disappointing, especially since some important early discoveries were made there (e.g., carbon nanotubes observed but not identified in electron micrographs in 1952), and despite the vast sums poured into the RusNano initiative (nominally at least, about $10,000 million since its creation in 2007–2008). China is apparently doing better than India (although a citation analysis might change this assessment). Korea's success seems to be strongly dependent on government rather than private funding, again apparently belying the analysis in Section 11.2, but there is no space here to carry out the more thorough analysis that would be needed to draw firm conclusions. At any rate the sums are impressive: $600 million has just been put into graphene research, development, and innovation there, compared with about $80 million in the UK (the National Graphene Institute currently being built in Manchester). Yet some smaller countries have more nanotechnology companies than South Korea—Canada has 70 and Australia has 41 ("nanowerk" website). Although nanotechnology is open to the participation of all countries, the output of many smaller ones is rather dismal. With a population of just under 10 million, Hungary only appears to have three nanotechnology companies, although the website fails to list the Research Institute for Technical Physics and Materials Science (MFA) of the Hungarian Academy of Sciences, which has a very active Nanostructures Department. Estonia, with a population of only 1.3 million, also has three nanotechnology companies listed, despite not having a single research organization involved in nanotechnology. For comparison, the state of Wyoming (population just above half a million) has three nanotechnology companies too (and one university involved in nanotechnology research).

14.1 Locating research partners

The foremost countries are of course those included in Figure 14.1, to which Switzerland should be added because of its intense research and spin-out activity. Within those countries, it is not necessarily the case that a handful of institutions are producing the overwhelming bulk of the exploratory research; important innovations may be widespread [1]. Sheer size plays a rôle; a small university may only achieve international prominence in one small niche area. Where spin-out activity is strong, as in Switzerland, the technology may already be in the commercial realm by the time it gets reported and an end-user may find it more expedient to collaborate with the spin-out company rather than the university research group in which the technology initially originated. Many micro companies have a single highly innovative product protected by a key patent.

The general level of technological infrastructure plays a vital role beyond the lowest levels of technological achievement. For this reason Russia is not a prominent player because, although new ideas might still be originating there, it is not possible to take them very far without overseas collaboration. The biggest unknown quantity is China (that is, wholly Chinese enterprises rather than joint ventures or subsidiaries of foreign companies). There is undoubtedly a tremendous volume of nanotechnology work going on there at both research and development levels [2], but it is less certain whether reliable supply partners can be found.

The most prolific institutions in the world are umbrella organizations comprising many individual institutes: the Chinese Academy of Sciences, the Russian Academy of Sciences, the Centre National de la Recherche Scientifique (CNRS) in France, in places 1, 2, and 3, respectively, for producing the most nanotechnology papers in 2005 [2]. France is relatively outstanding: the comparable Consiglio Nazionale delle Ricerche (CNR) in Italy only produces about two thirds of the output (and appears in rank 13). Position 4 is occupied by Tsing Hua University in China, and the next three positions by Japanese universities (Tohoku, Tokyo, and Osaka). The list is indeed dominated by the Far East. The most prolific institution in the USA is the University of Illinois (rank 17); there are only two other US universities (Berkeley and Texas) in the first 30. The only European university is Cambridge (rank 24); the Indian Institutes of Technology (another umbrella organization) only appear at rank 28. Recently the Rensselaer Polytechnic Institute (NY) has reported several significant discoveries in areas likely to be of interest to product developers.

Nowadays, with an unprecedented level of global networking and the highly efficient dissemination of knowledge (thanks to excellent communications), it is not a difficult problem to find a laboratory capable of carrying out almost any kind of nanotechnology research. Many, many universities and academic institutions are carrying out some work on nanotechnology. Facilities might be a limiting factor—even today, no single institution appears to have all that is required to undertake any kind of nanotechnology work, but if money is no obstacle equipment can be acquired or built, and specialized technical expertise is mobile [3].

Within the UK, the obvious (private) centers of excellence for nanotechnology include the BAE Systems Advanced Technology Centre (ATC) at Filton and the Qinetiq Advanced Microsystems Engineering at Malvern. Many universities are conducting some kind of nanotechnology research; the top four universities with respect to patenting activity are Cambridge, Oxford, Glasgow, and Imperial College (London).

Globally, leading nanotechnology companies (i.e., large companies that have made a serious commitment to nanotechnology, which may nevertheless only represent a very small part of their current product portfolio) are typically military (e.g., Raytheon, Lockheed Martin) or IT (e.g., IBM) in the USA. In Europe, chemical companies are very active in the development of nanomaterials (e.g., BASF in Germany). In Japan, engineering, electronics, and chemical companies are all actively carrying out nanotechnology research (e.g., Toyota, Toshiba, Mitsubishi).

14.2 Locating supply partners

Although the nanotechnology research community is very healthy, judging by the tremendous amount of publishing activity (one might even say overhealthy, because there is insufficient coordination, leading to duplicated effort and the neglect of gaps), the same cannot be said of the nanotechnology industry. There are numerous companies offering nanomaterials (overwhelmingly nanoparticles), mostly in the USA and China, but these materials are in no sense standardized and buyers are likely to have to pay a premium price because of the difficulty of comparison (absence of real competition). A highly significant milestone has been the launching, at the end of 2010, of the Integrated Nano-Science and Commodity Exchange (INSCX) [4], based in the UK but active globally, which offers trading services in nanomaterials and nanodevices akin to the well established exchanges for commodities such as metals and grains (cf. Section 11.3). This exchange is expected to transform the way nanomaterials are sourced. Among other effects, it is likely that the price of many nanomaterials will fall to a level at which their use can be contemplated (whereas at present, although there may be a clear technical advantage in employing them, they are simply too expensive).

14.3 Categories of countries

According to the Civilization Index (CI) [5], the countries of the world can be grouped into four categories:

I. High per-capita income, high level of scientific activity (e.g., Canada, Germany, Japan, Switzerland, UK, USA).
II. Low per-capita income, high level of scientific activity (e.g., Argentina, China, Georgia, Hungary, India, Russia).
III. High per-capita income, low level of scientific activity (e.g., Brunei, Kuwait, Libya, Saudi Arabia).
IV. Low per-capita income, low level of scientific activity (e.g., Angola, Indonesia, Thailand, Zambia).

Category I comprises the wealthiest countries of the world. They are active in nanotechnology, have a high level of scientific research and technical development in most areas, and have a good level of higher education. We would expect that countries in this category are leading in at least one branch, both scientifically and in developing innovative products.

Category II mostly comprises countries of the former Soviet Union, which had highly developed scientific research activities for much of the twentieth century but since 1991 have fallen on hard times, together with countries that historically had strong traditions of technical innovation (for example, until around the 17th century

China was well ahead of Europe) but failed to sustain past momentum (for reasons that are not understood) and, perhaps more significantly, failed to develop a strong science to parallel their technology; this subgroup within the category also has a large rural, barely educated population.

Category III comprises countries with a lower level of civilization that have acquired vast riches in recent decades through the export of raw materials found within their territories, notably mineral oil. They have manifested little interest in supporting the global scientific community; what technology they have is mostly imported.

Category IV comprises countries with a lower level of civilization that might require centuries of development before reaching the attainments of Category I (see Section 14.3.1).

Essentially all the nanotechnology-active countries are in Category I.

14.3.1 Nanotechnology in the developing world

An oft-debated issue is whether developing economies (the "Third World") can disproportionately benefit from adopting nanotechnology in order to shortcut the laborious path of technical development that has been followed by the "old" economies. A key idea is that for many products much less capital investment is required to set up nanomanufacturing than to set up conventional production (computer hardware need not be included in the manufacturing portfolio because high-performance chips are anyway available practically as commodities nowadays).

In other words, nanotechnology is *prima facie* very attractive for poor, technology-poor countries to embrace because it seems to require less investment before yielding returns (other than nanoscale semiconductor processing). Furthermore, nanotechnology offers more appropriate solutions to current needs than some of the sophisticated Western technologies available for import. Water purification using sunlight-irradiated titanium dioxide nanoparticles would be a characteristic example. Are these propositions reasonable?

Alas, the answer probably has to be "no." One of the greatest handicaps countries in Category IV face is appallingly ineffectual government; precisely where direction would be needed to focus local talent there is none, and most of these governments are mired in seemingly ineradicable venality. The situation nowadays in many African countries is apparently considerably worse than half a century ago when, freshly independent, they were ruled with enthusiasm and a great desire to develop a worthy autonomy. Zimbabwe offers a very sobering illustration: the country had a good legacy of physical and educational infrastructure from the colonial era, but today, after the government has bent over backwards to distribute land to the landless, they have shown themselves incapable of stewardship and agricultural output has plummeted.

The doubtless well-meaning efforts that have resulted in the foundation of institutions such as the new Library of Alexandria and the Academy of the Third World also seem doomed to failure, for they are rooted in an uncritical admiration of sterile "pure" sciences in the Western tradition which, while superficially glamorous in a narrow academic sense, are incapable of taking root and growing—nor would such growth be useful to their environment.

Having said that, if a country wished to focus all its resources on one area, nanotechnology would probably be the best choice, because its interdisciplinary nature would ensure that the knowledge base had to be broad, while the immediacy of applications would ensure rapid returns. The criterion of success will be for a country to achieve leadership in some subfield of the technology: this will show it has crossed the sustainability threshold, which is unlikely to be achievable merely by imitating leading Western scholarship.

An attractive feature of promoting "nanotechnology in the jungle," as we might call it, is its potential to benefit the overall economy through the promotion of disequilibrium as advocated by Hirschmann [6]: the technological demands of having any success at all in an advanced system of production necessarily forces the rest of the economy (parts of which must inevitably supply the ultrahigh technology part) to develop in its train.

It would be greatly encouraging if any country launching a focused nanotechnology effort would avoid the pitfalls of "standard empiricism" (see Chapter 17) and make a fresh start with aim-oriented science, and from the beginning encourage healthy, open criticism. At the same time, good use should be made of the global scientific legacy, by sending scholars to a variety of foreign centers of excellence to learn. It would be futile to await handouts from international funds (IMF, World Bank, and the like) for such a purpose—they are not interested in promoting independent science. In most countries, the leaders could well afford to fund appropriate scholarships for undertaking doctoral degrees (for example) abroad. Upon their return these scholars would be seeds of immense growth potential.

Despite their problems, these countries have two great advantages compared with the developed world. One is that they have practically nothing to dismantle first, which represents such a big obstacle to the introduction of new ways of thinking [7]. The other is that their natural resources are relatively unexplored and unexploited; looking at them from the bottom up is almost certain to yield new knowledge, leading to new avenues for wealth creation.

References and notes

[1] It is a prominent feature of the US National Nanotechnology Initiative that funds are distributed to enable "worthy but laggard" institutions to catch up with their peers.

[2] Kostoff RN et al. The growth of nanotechnology literature. Nanotechnol Perceptions 2006;2:229–47.

[3] As an extreme example, in 2009 an entire research-intensive university was created *ex nihilo* on a desert shore where only a few fishing villages existed beforehand (the King Abdullah University of Science and Technology, KAUST). It is, however, too early to assess the success of this venture. Although nanotechnology is one of its six priority areas, it remains to be seen what impact it will have on the global community.

[4] http://inscx.com.

[5] A new index for assessing development potential. Basel: Collegium Basilea; 2008.
[6] Hirschmann AO. The strategy of economic development. Yale: University Press; 1958.
[7] Cf. the introduction of mobile telephony: penetration exceeds that in the developed world, because the fixed line infrastructure is so poor.

Further reading

[1] OECD/NNI International symposium on assessing the economic impact of nanotechnology. Background Paper 1: challenges for governments in evaluating return on investment from nanotechnology and its broader economic impact. Paris: OECD (Working Party on Nanotechnology) (16 March 2012).

CHAPTER

Design of Nanotechnology Products

15

CHAPTER OUTLINE HEAD

15.1 The Challenge of Vastification 177
15.2 Enhancing Traditional Design Routes 178
15.3 Crowdsourcing [7] 179
15.4 Materials Selection 180

It has already been stressed in Chapter 10 that one of the difficulties faced by suppliers of any upstream technology is that they must ensure that its use is already envisaged in the design of the downstream products that will incorporate the technology. Apart from fulfilling technical specifications, esthetic design is furthermore one of the crucial factors determining the allure of almost any product (perhaps those destined for outer space are an exception), but especially a consumer one. In this chapter we look at some peculiar features associated with design of nanodevices, here defined as devices incorporating nanomaterials.

15.1 The challenge of vastification

There may be little point in making something very small if only a few of those things are required [1]. Interest in making a very large-scale integrated circuit with nanoscale components is rooted in the possibility of making vast numbers in parallel. Alongside diminishing component sizes, the diameter of the silicon wafers on which they are fabricated has grown from 4″ to 8″ to 12″.

Hence, although the most obvious consequence of nanotechnology is the creation of very small objects, an immediate corollary is that there will likely be a great many of these objects. If r is the relative device size and R the number of devices, then usefulness may require that $rR \sim 1$, implying the need for making 10^9 nanodevices at a stroke [2]. This corresponds to the number of components (with a minimum feature size of 45–65 nm) on a very large-scale integrated electronic processor or storage chip, for example. At present, all these components are explicitly designed

Applied Nanotechnology, Second Edition. http://dx.doi.org/10.1016/B978-1-4557-3189-3.00015-4

and fabricated. But will this still be practicable if the number of components increases by further orders of magnitude?

15.2 Enhancing traditional design routes

Regarding processor chips, which are presently the most vastified objects in the nano world, aspects requiring special attention are: power management, especially to control leakage; process variability, which may require a new conception of architectural features; and a systems-oriented approach, integrating functions and constraints, rather than considering the performance of individual transistors. Nevertheless, the basic framework remains the same.

Because it is not possible to give a clear affirmative answer to the above question, alternative routes to the design and fabrication of such vast numbers are being explored. The human brain serves as an inspiration here. Its scale is far vaster than the integrated circuit: it has $\sim 10^{11}$ neurons, and each neuron has hundreds or thousands of connections to other neurons. So vast is this complexity there is insufficient information contained in our genes to specify all these interconnections. We may therefore infer that our genes specify an algorithm for generating them [3].

In this spirit, evolutionary design principles may become essential for designing nanodevices. An example of an evolutionary design algorithm is shown in Figure 15.1. It might be initialized by a collection of existing designs, or guesses at possible new designs. Since new variety within the design population is generated randomly, the algorithm effectively expands the imagination of the human designer.

Although this strategy enables the design size (i.e., the number of individual features that must be explicitly specified) to be expanded practically without limit, one typically sacrifices knowledge of the exact internal workings of the device, introducing a level of unpredictability into device performance that may require a new engineering paradigm to be made acceptable.

Genetic algorithms [4] use bit strings to encode the target object. The genome is fixed in advance, only combinations of presence and absence of individual features can evolve. In other words, the form of the solution is predetermined. For example, if the solution can be expressed as an equation, the coefficients evolve but not the form of the equation. More advanced algorithms relax these conditions; that is, the genome length can vary (i.e., gene additions and deletions are possible). These schemata are rather far from natural selection, and might best be described as artificial selection.

Genetic programming [5] works at a higher level, in which the algorithm itself evolves. In other words, the form of the solution can evolve. Typically the solution is defined by trees of Lisp-like expressions, and changes can be made to any node of the tree. Genetic programming is closer to natural selection.

Human knowledge can be captured not only in the design of the algorithms, but also by incorporating an interactive stage in the fitness evaluation [6].

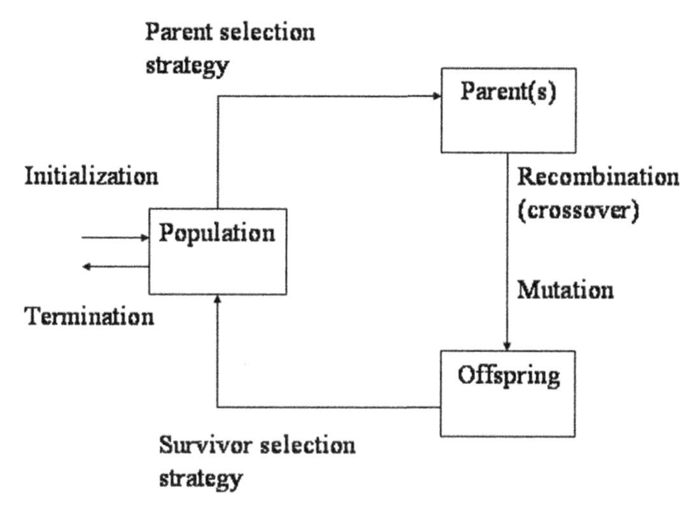

FIGURE 15.1

An evolutionary design algorithm. All relevant design features are encoded in the genome (a very simple genome has each gene represented by a single binary digit indicating absence (0) or presence (1) of a feature). The genomes are evaluated ("survivor selection strategy")—this stage could include human (interactive) as well as automated evaluation—and only genomes fulfilling the evaluation criteria are retained. The diminished population is then expanded in numbers and in variety—typically the successful genomes are used as the basis for generating new ones via biologically inspired processes such as recombination and mutation.

15.3 Crowdsourcing [7]

The stupendous increase in global connectivity between individuals has spawned a completely new paradigm for knowledge exchange. It is most obviously applicable to goods that can be transferred by the internet, such as design and software, but since almost everything can be adequately encoded as a binary sequence of symbols (including the instructions to manufacture a physical object) it does not have any real limitation. "Outsourcing" is the distribution of defined tasks to people working in locations remote from that of the originator, but crowdsourcing goes well beyond that, because typically only the final goal is specified; the specifier hopes to receive a vast variety of ideas that will solve, or help to solve, the problem.

In principle anyone could solicit contributions; several organizations have been created to facilitate the process, giving more publicity to the demand and constructing a mechanism for rewarding contributors whose ideas are taken up. Ultimately that might be seen as superfluous. We have the examples of open source software (e.g., Linux) and the enormously successful open source encyclopedia Wikipedia—just contrast the rise of the latter with the formerly mighty Encyclopaedia Britannica (which was, admittedly, anyway in decline since its great 11th edition published in 1910–11), in which contributions are provided *gratis*. Productive Nanosystems (Section 16.1) will doubtless see a similar growth of open source instruction sets for nanofabricating artifacts.

A related development is "unsourcing", the reliance on contributions from internet volunteers for technical support to proprietary products. The proprietor may have to provide some initial impetus by setting up an online community. Some companies have sought to incentivize the process by turning the process into a game ("gamification").

Clearly these developments have vast implications for the patent system for protecting intellectual property (cf. Section 10.11). It will, in fact, become quite redundant. Although it might be hard to envisage the dismantling of the huge and elaborate apparatus that has been built up around patents during the past few hundred years, society probably has to get used to the idea that it is going to fade away. The notion of a kind of copyright for ideas will likely be replaced by a kind of collective ownership. The digital age (of which nanotechnology is the material embodiment) converts everything into ideas.

15.4 Materials selection

Ashby has systematized materials selection through his property charts [8]. For example, Young's modulus E is plotted against density ρ for all known materials, ranging from weak light polymer foams to strong dense engineering alloys. The huge interest in nanomaterials is that it may be possible to populate empty regions on such charts, such as strong and light (currently natural woods are as close as we can get to this) or weak and dense (no known materials exist).

Material properties are only the first step. Shapability is also important, in ways that cannot be easily quantified. For example, rubber can readily be manufactured as a sealed tube in which form it can serve as a pneumatic tyre, but it is at risk from punctures, and a novel solid material may be useful and more robust for the same function. Finally, availability (including necessary human expertise) and cost—linked by the laws of supply and demand—must be taken into consideration. Nanotechnology, by allowing rapid material prototyping, should greatly enhance the real availability of novelty. An assembler should in principle allow any combination of atoms to be put together in some way to create new materials.

References and notes

[1] Devices for which accessibility is the principal consideration might still be worth making very small even if only few are required; e.g., for a mission to outer space.

[2] This is why vastification—proliferation of vast numbers of objects—almost always accompanies nanification.

[3] Érdi P, Barna Gy. Self-organizing mechanism for the formation of ordered neural mappings. Biol Cybernetics 51 (1984) 93–101.

[4] Holland JH. Adaptation in natural and artificial systems. Ann Arbor: University of Michigan Press; 1975.
[5] Koza JH. Genetic programming. Cambridge, MA: MIT Press; 1992.
[6] E.g., Brintrup AM et al. Evaluation of sequential, multi-objective, and parallel interactive genetic algorithms for multi-objective optimization problems. J Biol Phys Chem 6 (2006) 137–46.
[7] References (e.g., to web sites) are not given in this section because the field is evolving so rapidly.
[8] Ashby MF. Materials selection in mechanical design. Oxford: Pergamon; 1992.

Further reading

[1] Banzhaf W, Beslon G, Christensen S, Foster JA, Képès F, Lefort V, Miller JF, Radman M, Ramsden JJ. From artificial evolution to computational evolution: a research agenda. Nature Rev Genet 2006;7:729–35.

PART

WIDER AND LONGER-TERM ISSUES

CHAPTER

The Future of Nanotechnology

16

CHAPTER OUTLINE HEAD

16.1 Productive Nanosystems 186
16.1.1 The Technology 187
16.1.2 Social Impacts 188
16.1.3 Timescales 189
16.2 Self-Assembly 190
16.3 Molecular Electronics 191
16.4 Quantum Computing 192

Whereas most of this book has been devoted to the prediction of substitutional and incremental nanotechnology, in this chapter we address the long-term future, for which the traditional methods of economic forecasting are of little use.

As pointed out by Toth-Fejel [1], important ways to predict the future include:

Via prophets, who are individuals with charisma, a track record of successful predictions (ideally based on an intelligible chain of reasoning) and the courage to contradict popular opinion. The value of the prophet's work might be primarily derived from a cogent marshaling of relevant data; the prophecy is important when it is accompanied by a similar creative leap as when a theory emerges from a mass of experimental data.

Through history: one looks for patterns in the past to find analogies for, or extrapolations into, the future. Predictions tend to be necessarily rather vague—that is, made at a fairly high level. This method does not have a good track record, despite significant apparent successes (e.g., the First World War following from the Franco-Prussian War because the latter's terms of peace were too onerous for France, and the Second World War following from the First World War because the latter's terms of peace were too onerous for Germany; doubtless some participants of the Versailles peace conferences were aware of the dangers of what was being done, but the proceedings got bogged down in a morass of detail and were bedeviled by partisan considerations).

Trend ranking: one reasonably assumes that the significance of a trend depends on its rate of change and duration: Typically, highly significant trends (e.g., accelerating

Applied Nanotechnology, Second Edition. http://dx.doi.org/10.1016/B978-1-4557-3189-3.00016-6

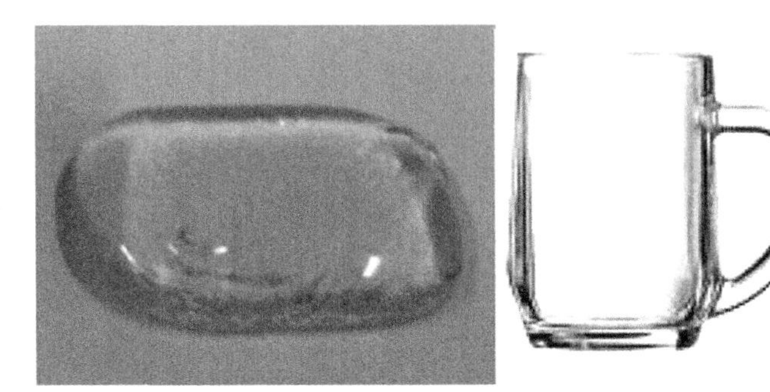

FIGURE 16.1

(Left) Final state of a mass of molten glass lead on the table; (right) final state of a mass of molten glass after application of the necessary external operations to produce a mug. Reproduced from B. Laforge, Emergent properties in biological systems as a result of competition between internal and external dynamics. *J. Biol. Phys. Chem.* 9 (20) 5–9 with permission of Collegium Basilea.

technology, increasing recognition of human rights) will enslave weaker ones (slow and short, e.g., business cycles and fashion).

Engineering vs science: scientific discoveries (e.g., X-rays, penicillin, Teflon) are impossible to predict (we exclude discoveries of facts (e.g., the planet Uranus) that were predicted by theories, in the formulation of which previously discovered facts played a rôle). On the other hand, engineering achievements (e.g., landing a man on the Moon) are predictable applications of existing knowledge that adequate money and manpower solved on schedule. According to the new model (Figure 2.3), new technologies (e.g., atomic energy and nanotechnology) are closely related to scientific discovery, making them concomitantly harder to predict.

The will to shape the future: the idea that the future lies in man's hands (i.e., he has the power to determine it) [2]. This stands in direct opposition to predestination. Reality is, of course, a mixture of both: the future involves unpredictability but is subject to certain constraints (see Figure 16.1 for an illustration).

Scenarios: not included in Toth-Fejel's list, but nevertheless of growing importance; e.g., in predicting climate change [3].

16.1 Productive nanosystems

The technological leap that is under consideration here is the introduction of desktop personal nanofactories [4]. These are general-purpose assemblers that represent the ultimate consummation of Richard Feynman's vision, capable of assembling things atom-by-atom using a simple molecular feedstock such as acetylene or propane, piped into private houses using the same kind of utility connection that today delivers natural gas. Such is the nature of this technology that once one personal nanofactory is introduced, the technology will rapidly spread, certainly throughout the developed world. It may be assumed that almost every household will purchase one [5]. What, then, are the implications of this?

16.1.1 The technology

Science of this era of Productive Nanosystems can be summed up as a quasi-universal system of "localized, individualized ultralow-cost production on demand using a carbon-based feedstock." Let us briefly take each of these attributes in turn.

Localized production will practically eliminate the need for transport of goods. Transport of goods and people accounts for 28% of fossil fuel usage (compared with 32% used by industry) [6], possibly 90% of which would no longer be necessary with the widespread introduction of the personal nanofactory. This would obviously have a hugely beneficial environmental impact (cf. Section 7.4).

We have become accustomed to the efficiency of vast central installations for electricity generation and sewage treatment, and even of healthcare, but future nanotechnology based on Productive Nanosystems will reverse that trend. Ultimately it will overturn the paradigm of the division of labor that was such a powerful concept in Adam Smith's conception of economics. In turn, globalization will become irrelevant and, by eliminating it, one of the gravest threats to the survival of humanity, due to the concomitant loss of diversity of thought and technique, will be neutralized.

Individualized production or "customized mass production" will be a powerful antidote to the products of the Industrial Revolution that are based on identical replication. In the past, "to copy" (e.g., a piece of music) meant copying it out by hand from a preexisting version. This was in itself a powerful way of learning for past generations of music students. Nowadays it means making an identical photocopy using a machine. In the Roman empire, although crockery was made on a large scale, each plate had an individual shape; almost two millennia later, Josiah Wedgwood rejoiced when he could make large numbers of identical copies of one design. The owner of a personal nanofactory (the concrete embodiment of a Productive Nanosystem) will be able to program it as he or she wishes (as well as having the choice of using someone else's design software).

Ultralow-cost production will usher in an era of economics of abundance. Traditional economics, rooted in the laws of supply and demand, are based on scarcity. The whole basis of value and business opportunities will need to be rethought.

Production on demand also represents a new revolutionary paradigm for the bulk of the economy. Only in a few cases—the most prominent being Toyota's "just-in-time" organization of manufacture—has it been adopted in a significant way. A smaller-scale example is provided by the Italian clothing company Benetton—undyed garments are stored centrally, and dyed and shipped in small quantities according to feedback regarding what is selling well from individual shops—and a similar approach is adopted by the Spanish clothing company Zara. Not only does this lead to a reduction of waste (unwanted production), but also to elimination of a significant demand for credit, which comes from production in anticipation of demand. Personal nanofactory-enabled production on demand represents the apotheosis of these trends.

Carbon-based feedstock. The implications of carbon-based feedstock (acetylene or propane, for example) as a universal fabrication material are interesting. The production of cement, iron and steel, glass and silicon account for about 5% of global carbon emissions. Much of this would be eliminated. Furthermore, the supply of feedstock could, given an adequate supply of energy, be sequestered directly from the atmosphere.

16.1.2 Social impacts

Although the anticipated course of nanotechnology-based technical development can be traced out, albeit with gaps, and on that basis a fairly detailed economic analysis carried out [7], ideas regarding the social impacts of these revolutionary changes in manufacturing are far vaguer. An attempt was made a few years ago [8], (typically) stating that "nanotechnology is being heralded as the new technological revolution … its potential is clear and fundamental … so profound that it will touch all aspects of the economy and society. Technological optimists look forward to a world transformed for the better by nanotechnology. For them it will cheapen the production of all goods and services, permit the development of new products and self-assembly modes of production, and allow the further miniaturization of control systems. They see these effects as an inherent part of its revolutionary characteristics. In this nano society, energy will be clean and abundant, the environment will have been repaired to a pristine state, and any kind of material artifact can be made for almost no cost. Space travel will be cheap and easy, disease will be a thing of the past, and we can all expect to live for a thousand years" [9]. Furthermore, these writings remain silent about how people will *think* under this new régime; their focus is almost exclusively on material aspects. There is perhaps more recognition of nanotechnology's potential in China, where the Academy of Sciences notes that "nanodevices are of special strategic significance, as they are expected to play a critical role in socio-economic progress, national security and science and technology development."

Traditional technology (of the Industrial Revolution) has become something big and powerful, tending to suppress human individuality; men must serve the machine. Moreover, much traditional technology exacerbates conflict between subgroups of humanity. This is manifested not only in the devastation of vast territories by certain extractive industries, but also by the "scorched earth" bombing of cities such as Dresden and Hamburg in World War II.

In contrast, nanotechnology is small without being weak and is perhaps "beautiful." Since in its ultimate embodiment as Productive Nanosystems it becomes individually shapable, it does not have all the undesirable features of "big" technology; every individual can be empowered to the degree of his or her personal interests and abilities. It is therefore important that in our present intermediate state nanotechnology is not used to disempower [10].

Possibly thanks to that rather influential book by E.F. Schumacher, "Small Is Beautiful" (1973), nanotechnology started with a generally benevolent gaze cast upon it. The contrast is especially striking in comparison with biotechnology, one of the recent products of which, namely genetically modified crops, has excited a great deal of controversy that is far from being settled (and probably cannot be without more extensive knowledge of the matter, particularly at the level of ecosystems). Wherever bulk is unnecessary, miniaturization must necessarily be good, and what else is nanotechnology if not the apotheosis, in any practical sense, of miniaturization? This appearance of a favorable public opinion must, however, be tempered by the knowledge that only a few percent of the population actually have an intelligible notion of what nanotechnology is. One might expect that the more solid the tradition of scientific journalism, the higher the percentage—the Swiss are rather well informed—but in France, although having such a tradition almost as good as in Switzerland, the percentage of the general public that knows something about nanotechnology seems to be no greater than in the UK. The English language press of the world does not, unfortunately, seem to have set itself a very high standard for nanotechnology reporting. Interest in the topic enjoyed a brief surge after the publication of Michael Crichton's novel "Prey" (2000), but has not persisted. There has been considerable effort, some of it sponsored by the state, to disseminate knowledge about nanotechnology among schoolchildren, with the result that they are possibly the best-informed section of the population.

The potential of nanotechnology is surely positive, because it offers the opportunity for all to fully participate in society. The answer to the question how one can move more resolutely in that direction would surely be that under the impetus of gradually increasing technical literacy in an era of leisure, in which people are as much producers as consumers, there will be a gradually increasing level of civilization, including a more profound understanding of nature. The latter in particular must inevitably lead to revulsion against actions that destroy nature, and that surely is how the environment will come to be preserved. In fact, the mission of the Field Studies Council (FSC), which was born around the time of the UK 1944 Education Act, namely "Environmental Understanding for All," is a worthy exemplar for the nano-era, which could have as its mission "Technical Understanding for All." And, just as the FSC promoted nature conservancy [11], it is appropriate for nanotechnology, with its very broad reach into all aspects of civilization, to have a wider mission, namely "Elevation of Society." This, by the way, implies other concomitant advances, such as in the early (preschool) education of infants, which has nothing to do *per se* with nanotechnology, but which will doubtless be of crucial importance in determining whether humanity survives.

16.1.3 Timescales

"True" or "hardcore" nanotechnologists assert that the goal of nanotechnology is Productive Nanosystems, and that the question is "when," not "if." Opponents implicitly accept the future reality of assemblers, and oppose the technology on the grounds of the dangers (especially that of "gray goo"—assemblers that run out of control and do nothing but replicate themselves, ultimately sequestering the entire resources of the Earth for that purpose). Finally there is a group that asserts that nanotechnology is little more than nanoparticles and scanning probe microscopes and that all the fuss, even the word "nanotechnology," will have evaporated in less than a decade from now.

This last attitude is rather like viewing the Stockton and Darlington Railway as the zenith of a trend in transportation that would soon succumb to competition from turbocharged horses. And yet, just as the company assembled on the occasion of the Rainhill engine trials could have had no conscious vision of the subsequent sophistication of steam locomotives such as Caerphilly Castle, Flying Scotsman, or Evening Star (not to mention those designed in France by André Chapelon, which were even more advanced) and would have been nonplussed if asked to estimate the dates when machines fulfilling their specifications would be built, so it seems unreasonable to demand a strict timetable for the development of advanced nanotechnology. It should be emphasized that by the criterion of atomically precise manufacturing, today's nanotechnology—overwhelmingly nanoparticles—is extremely crude. But this is only the first stage, that of passive approximate nanostructures. Applications such as sunscreen do not require greater precision. Future envisaged phases are:

Active nanodevices able to change state, transform and store information and energy, and respond predictably to stimuli. Integrated circuits with 65 nm features (made by "top-down" methods) belong here. Nanostorage devices (e.g., based on single electrons or molecules), biotransducers, and the quantum dot laser are examples that have reached the stage of proof of principle. It is noteworthy that self-assembly ("bottom–up") nanofacture is being pursued for some of these.

Complex machines able to implement *error correction codes*, which are expected to improve the reliability of molecular manufacturing by many orders of magnitude (consider chemical syntheses with error rates around 1 in a 100: a yield of 99% is considered outstanding); natural protein synthesis with error rates of 1 in 10^3-10^4; DNA with error rates of 1 in 10^6; and modern computers have error rates better than 1 in 10^{23} operations thanks to error detection and correction codes originally developed by Hamming and others, without which pervasive low-cost computing and all that depends on it, such as the internet, would not be possible. Algorithmic concepts are very significantly ahead of the physical realization (see Section 1.1). The main practical approaches currently being explored are tip-based nanofabrication (i.e., diamondoid mechanosynthesis; or patterned depassivation followed by atomic layer epitaxy) and biomimicry (DNA "origami" and bis-peptide synthesis).

Productive Nanosystems, able to make atomically precise tools for making other (and better) Productive Nanosystems, as well as useful products. Current progress and parallels with Moore's law suggest that they might be available in 10–20 years.

16.2 Self-assembly

Although "passive" self-assembly creates objects of indeterminate size (except in the special case of competing interactions of different ranges [12]) and, hence, is not

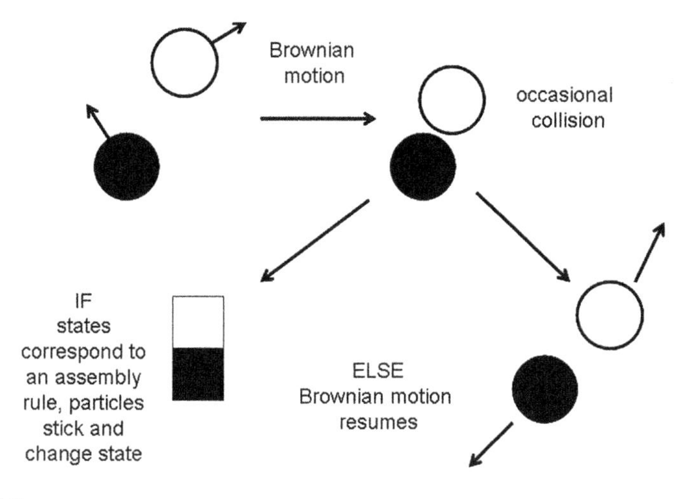

FIGURE 16.2

Illustration of programmable self-assembly.

useful for most technological applications (especially device nanofacture), biology shows that useful in self-assembly is possible (e.g., the final stages of assembly of bacteriophage viruses [13]). It depends on initial interactions altering the conformations of the interacting partners and, hence, the spectrum of their affinities (see Figure 16.2). Called programmable self-assembly (PSA), it can be formally modeled by graph grammar, which can be thought of as a set of rules encapsulating the outcomes of interactions between the particles [14]. While macroscopic realizations of PSA have been achieved with robots, we seem to be a long way off mimicking biological PSA in wholly artificial systems. Modifying biological systems is likely to be more achievable, and there is intensive research in the field [15]. The approach seems to be at least as promising as the assembler concept.

16.3 Molecular electronics

The industry view of the continuation of Moore's law is supposed to be guaranteed for several more years via further miniaturization and novel transistor architectures. Another approach to ultraminiaturize electronic components is to base them on single organic molecules uniting an electron donor (D^+, a cation) and acceptor (A^-, an anion) separated by an electron-conducting bridge (a π-conjugated (alkene) chain). The molecule is placed between a pair of (usually dissimilar) metal electrodes $M^{(1)}$ and $M^{(2)}$ [16], chosen for having suitable work functions and mimicking a semiconductor p–n junction. Forward bias results in $M^{(1)}/D^+ - \pi - A^-/M^{(2)} \rightarrow M^{(1)}/D^0 - \pi - A^0/M^{(2)}$, followed by intramolecular tunneling to regenerate the starting state. Reverse bias tries to create $D^{2+} - \pi - A^{2-}$, but this is energetically unfavorable and hence electron flow is blocked (rectification). This technology is still in the research phase, with intensive effort devoted to increasing the rectification ratio.

16.4 Quantum computing

Extrapolation of Moore's law to about the year 2020 indicates that component size will be sufficiently small for the behavior of electrons within them to be perturbed by quantum effects, implying the end of the semiconductor road map and conventional circuit logic. Another problem with logic based on moving charge around is energy dissipation. Quantum logic (based on superposition and entanglement) enables computational devices to be created without these limitations and intensive academic research is presently devoted to its realization.

The physical embodiment of a bit of information—called a qubit in quantum computation—can be any absolutely small object capable of possessing the two logical states 0 and 1 in superposition, e.g., an electron, a photon, or an atom. A single photon polarized horizontally (H) could encode the state $|0\rangle$ and polarized vertically (V) could encode the state $|1\rangle$ (using the Dirac notation). The photon can exist in an arbitrary superposition of these two states, represented as $a\,|\mathrm{H}\rangle + b\,|\mathrm{V}\rangle$, with $|a|^2 + |b|^2 = 1$. The states can be manipulated using birefringent waveplates, and polarizing beamsplitters are available for converting polarization to spatial location. With such common optical components, logic gates can be constructed [17]. Another possible embodiment of a qubit is electron spin (a "true" spintronics device encodes binary information as spin, in contrast to the so-called spin transistor, in which spin merely mediates switching) [18].

References and notes

[1] Toth-Fejel T. Irresistible forces vs immovable objects: when China develops productive nanosystems. Nanotechnol Perceptions 2008;4:113–32.

[2] One of the more prominent philosophers associated with this idea was F. Nietzsche. However, he also believed in the idea of eternal return (the endless repetition of history).

[3] Anissimov M et al. The center for responsible nanotechnology scenario project. Nanotechnol Perceptions 2008;4:51–64.

[4] Drexler KE. Nanosystems: molecular machinery, manufacturing, and computation. New York: Wiley; 1992.

[5] Freitas (*loc. cit.*) assumes a 20 year interval for their introduction.

[6] Holt GC, Ramsden JJ. Introduction to global warming. In: Ramsden JJ, Kervalishvili PJ, editors. Complexity and security. Amsterdam: IOS Press; 2008. p. 147–84.

[7] Freitas Jr RA. Economic impact of the personal nanofactory. Nanotechnol Perceptions 2006;2:111–26.

[8] Wood SJ et al. The social and economic challenges of nanotechnology. Swindon: Economic and Social Research Council; 2003.

[9] For a critique of this report, see Ramsden JJ. The music of the nanospheres. Nanotechnol. Perceptions 2005;1:53–64.

[10] An example of disempowerment is the recent development of "theranostics"—automated systems, possibly based on implanted nanodevices, able to autonomously diagnose disease and automatically take remedial action; for example by releasing drugs. In contrast to present medical practice, in which a practitioner diagnoses, perhaps imperfectly, and proposes a therapy, which the patient can accept or refuse, theranostics disempowers the patient, unless he was involved in writing the software controlling it.
[11] Berry RJ. Ethics, attitudes and environmental understanding for all. Field Studies 1993;8:245–55.
[12] Ramsden JJ. The stability of superspheres. Proc R Soc Lond A 1987;413:407–14.
[13] Kellenberger E. Assembly in biological systems. In: Polymerization in biological systems, CIBA foundation symposium 7 (new series). Amsterdam: Elsevier; 1972.
[14] Klavins E. Universal self-replication using graph grammars. In: International Conference on MEMs, NANO and smart systems, Banff, Canada; 2004.
[15] Chen J, Jonoska N, Rozenberg G, editors. Nanotechnology: science and computation. Berlin: Springer; 2006.
[16] See, e.g., Martin AS et al. Molecular rectifier. Phys Rev Lett 1993;70:218–21.
[17] Politi A, O'Brien JL. Quantum computation with photons. Nanotechnol Perceptions 2008;4:289–94.
[18] Bandyopadhyay S. Single spin devices—perpetuating Moore's law. Nanotechnol Perceptions 2007;3:159–63.

Further reading

[1] Allen PM. Complexity and identity: the evolution of collective self. In: Ramsden JJ, Aida S, Kakabadse A, editors. Spiritual motivation: new thinking for business and management. Palgrave Macmillan: Basingstoke; 2007. p. 50–73.
[2] Kim S. Directed molecular self-assembly: its applications to potential electronic materials. Electronic Materials Lett 2007;3:109–14.

CHAPTER

Society's Grand Challenges 17

CHAPTER OUTLINE HEAD

17.1 Material Crises . . . 195
17.2 Social Crises . . . 198
17.3 Is Science Itself in Crisis? . . . 198
17.4 Nanotechnology-Specific Challenges . . . 199
17.5 Globalization . . . 199
17.6 An Integrated Approach . . . 200

Human society is widely considered to have entered a difficult period. It is confronted with immense challenges of a globally pervasive nature. Extrapolating present trends leads to a grim picture of the possibility of a miserable collapse of civilization. Because of globalization, the collapse is likely to be global—whereas in the past, different "experiments" (types of socio-economic organization) were tried in different places, and the collapse of one (e.g., the Aztec empire) did not greatly affect others. During the previous half-century, destruction by nuclear weapons was considered to be the greatest threat, but this was clearly something exclusively in human hands, whereas now, even though the origins of the threats (climate change) might be anthropogenic, mankind seems to be practically powerless to influence the course of events.

Can a new technology help? Several decades ago, nuclear fusion—using the vast quantities of deuterium in the oceans as fuel—was seen as a way to solve the challenge of rapidly depleting fossil fuel reserves. As it happens, that technology has not delivered the promised result, but now a similar latent potential inheres in nanotechnology, which as a universal technology should in principle be able to resolve all the crises.

17.1 Material crises

The crises, and the potential contributions of nanotechnology, are:

Climate change, especially global warming [1]. If the cause is anthropogenic release of carbon dioxide into the atmosphere, then any technology that tends to diminish it

Applied Nanotechnology, Second Edition. http://dx.doi.org/10.1016/B978-1-4557-3189-3.00017-8

will be beneficial, and nanotechnology in the long term should have that capability (see Section 16.1). If the cause is not anthropogenic and due to (for example) variations in the solar constant, then we need to enhance our general technological capabilities to give us the power to combat the threat.

Demography. This comprises both population growth, considered to be excessive, and becoming too large for Earth to support, and aging of the population.

The latter is partly a social matter. It is customary for elderly people to retire from active work, but their income as *rentiers* (i.e., old-age pensioners) is ensured by those still actively working, so if the ratio of the latter to the former decreases, the pension system collapses (except in countries such as Switzerland where each worker invests in his or her future pension, rather than paying the pensions of those who are already retired). There is also the matter of healthcare: elderly people tend to require more resources. Advances in medical care, to which nanotechnology is making a direct contribution (Chapter 9), will diminish the healthcare problem. Therefore, elderly people should be able to continue to work longer, diminishing the threat to the pension system. At the same time, advances in technology should further diminish the prevalence of unpleasant jobs from which one is glad to retire. In any case, relatively unproductive work (e.g., involving an advisory or decision-making rôle without an impact on production—town planning is a good example, along with membership of the now numerous councils or committees with comparable functions) could be preferentially assigned to elderly people.

If, in principle at least, the problem of aging populations can thus be solved, the same cannot be said for population growth. It is probably best to consider it as a medical problem (as it is already in many countries), in which case the technology of Chapter 9 is applicable.

Environmental degradation. This is mostly a gradual change, but by being slow it is pernicious, and suddenly we have a seemingly irreparable dust bowl in the Mid-West or an Aral Sea disaster. The latter was a fairly direct result of the massive diversion of the two main feeder rivers, the Amu Darya and the Syr Darya, into irrigation, mainly of cotton fields, although unpredictable nonlinearities appeared near the end. Whether the immediate restoration of full flow would regenerate the sea is a moot point, and anyway has not been seriously considered because of the immense social dislocation that would result from the collapse of the cotton agro-industry. In Soviet times serious consideration *was* given to the possible solution of diverting some of the great, northward-flowing Siberian rivers into the Aral Sea, but the idea was ultimately abandoned due to fears of adverse ecological and climatic consequences. Some of the world's great deserts such as the Sahara and the Gobi are also currently expanding, but this may be a cyclic phenomenon linked to long-term climate changes. In fact, the mechanisms of desertification are not well understood, and efforts to investigate it piecemeal. Presumably the United Nations Organization declared 2007 as the Year of Desertification in an effort to improve matters, but it was singularly ineffectual in inspiring a coordinated global effort to tackle the problem. In any case, it is not clear how nanotechnology can contribute.

Another major challenge is environmental pollution. Certain remediation technologies based on nanoparticles (cf. Section 6.7) may help to alleviate local problems. In the long term, nanotechnology will certainly help to alleviate pollution via significantly increasing the overall efficiency of production (Chapter 16)—waste in manufacturing should be essentially eliminated. Nevertheless, artifacts will presumably still be used and discarded. Nanotechnology, however, offers the ultimate recycling technology—every discarded artifact can be decomposed into atoms, which are then sorted to be reused as feedstock.

Depletion of resources. Nanotechnology should have a generic beneficial effect, because if the same function can be achieved using less material, obviously fewer resources will be used.

Furthermore, very light yet strong materials (probably composites based on carbon nanotubes or graphene) are likely to be of great value for space travel—including the space elevator, which would enormously facilitate departures from Earth. Continuing increases in processing power will enhance the feasibility of unmanned space missions (enabling on-board manufacturing)—for example, to neighboring asteroids in order to mine metals no longer obtainable on Earth.

Financial chaos. The new economic order associated with nanotechnology—productive nanosystems (Chapter 16)—based on-production-on-demand may solve this problem automatically, since the rôle of credit will diminish. Given that the realization of productive nanosystems is not anticipated before at least another decade, and quite possibly two, have elapsed, it must not be hoped for that nanotechnology will come to the rescue of the present troubles, but perhaps it will prevent a fresh occurrence of bubbles.

Terrorism. This is above all a social problem [2], which might disappear if nanotechnology ushers in a new ethical era (see Chapter 18).

It is doubtful whether all the challenges can be met simultaneously, therefore priorities will have to be set. One notices that demography (especially population growth) is, in fact, the most fundamental challenge, in the sense that solving this one will automatically solve the others. One is dismayed that countries whose populations are falling (many European countries, Russia, and Japan) are being encouraged to promote immigration—to stave off collapse of their social security systems! The prolongation of healthy human life is one of the more reliably extrapolatable trends, and a clear corollary is that world population should stabilize at a lower level than otherwise. In blunt ecological terms, "be fruitful and multiply" is an appropriate injunction in a relatively empty world in which r-selection (see Section 3.1) operates, but in our present crowded, technologically advanced era K-selection is appropriate, typified by a sparser, but longer-living population.

17.2 Social crises

Material problems are usually in the forefront of attention, but any technological revolution also brings psychological and social problems in its wake. One of the general problems of technology increasing leisure (see Figure 2.2) is that people might find it harder to lead meaningful lives. We may have to ask whether we really need a further increase in the abundance of labor-saving devices. More attention will need to be paid by everybody to continue to exercise body and mind: Hebb's rule essentially guarantees that the brain will atrophy in the absence of thought trajectories. This issue, and related ones, is taken up more fully in the next and final chapter.

17.3 Is science itself in crisis?

The idea of scientific endeavor being harnessed to explicitly solve grave global problems is an attractive one. Bacon's stress on one of the purposes of scientific investigation being for "the relief of man's estate" would encourage that idea, and most scientists would probably agree that a basic humanitarian aim of science is to help promote human welfare. However, as Maxwell has pointed out, science seeks this by pursuing the purely intellectual aim of acquiring knowledge in a way (called standard empiricism (SE) by Maxwell) that is sharply dissociated from all consideration of human welfare and suffering [3]. Under the aegis of SE, any desire to solve global challenges is likely to be little more than a velleity. Maxwell advocates replacing SE by aim-oriented empiricism (AOE) as a better philosophy of science. Not only is it more rigorous, but value (to humanity and civilization) becomes an intrinsic part of its pursuit. Even in a highly abstract discipline the adoption of AOE will be a step forward, because of its greater rigor, although the fruits (research output) are not likely to look very different. In areas other than the "hard" sciences, the difference is likely to be dramatic. The social sciences, including sociology and economics, have become largely useless to humanity. On the contrary, astonishingly and sadly, the attitudes prevailing among many academic sociologists and economists tend to drag humanity down whenever they are taken up by politicians or other activists. Social science should be replaced by something called social inquiry, social methodology, or social philosophy, concerned to help humanity tackle its immense problems of living in more rational ways than at present, and seeking to build into social life progress-achieving methods arrived at by generalizing AOE, the progress-achieving methods of the natural sciences.

The natural sciences themselves need to acquire a tradition of criticism that has long been a part of literary and artistic work. Because, unlike those other areas of endeavor, the natural sciences contain their own internal validation—the predictions of a theory can always be tested via empirical observation—they have not felt the same need to develop a tradition of criticism that in the literary world is as esteemed as the creation of original works. In consequence, much science tends to move in sterile or even counterproductive directions [4]. One often hears reference to the "rapid progress" in many fields, especially the biological sciences. To be sure, advances in techniques and instrumentation have yielded impressive results, but if the direction is wrong, they do not mean very much. These weaknesses will become more obvious when the challenges, to which science as presently practised might be able to respond, but does not or cannot, become more important.

17.4 Nanotechnology-specific challenges

This section addresses the possibility of new, survival-threatening crises emerging from the growth of nanotechnology. Any revolution brings its own attendant new challenges (typically referred to as "birth pangs" and the like). There is already widespread implicit recognition of them. An example is presented by the report on nanoparticle risks commissioned by the British government [5]. This addresses the need to assess human exposure to engineered nanomaterials, evaluate their toxicity, and develop models to predict their long-term impacts. Similar investigations should be undertaken to establish effects on the overall ecosystem, including plant and microbial life. Given the extensive data that already exists, at least concerning human health impacts [6], care should be taken to avoid the waste and futility of endless studies aimed at the same goal, and all deficient in some regard. Meanwhile, more attention should be paid to how to act efficaciously upon the findings.

Productive Nanosystems (PNs, Section 16.1) raise the risk of "gray goo" (the uncontrolled subversion of all terrestrial matter to assembling assemblers). Given that PNs are not expected before at least 10–20 years from now, should we already be concerned at this eventuality? Probably not, since it is still associated with so many imponderables.

There is strong military interest in nanotechnology [7], raising the specter that it will "met en oeuvre des moyens de mort et de destruction incomparablement plus efficaces que par le passé" [8]. Albert Schweitzer points out that history shows that victory by no means always belongs to the superior civilization; as often as not a more barbaric power has conquered. This problem is not specially associated with nanotechnology; but there is at least the hope that a pre-singularity surge of human solidarity may neutralize it (see also Section 17.6 and Chapter 18).

17.5 Globalization

Perhaps *the* greatest socio-economic-technical danger faced by humanity is that of globalization. Advances in transport and communication technology have made it seem inevitable, and it appears as the apotheosis of Adam Smith's economic system (based on the division of labor) that has been so successful in augmenting the wealth of mankind. Yet globalization carries within it the seeds of great danger: that of *diminishing and fatally weakening the diversity that is so essential a part of our*

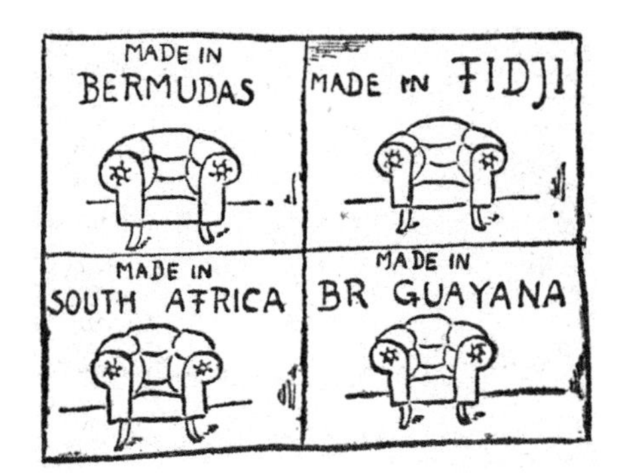

FIGURE 17.1

Exhibits at the 1924 British Empire Exhibition in London (Wembley), drawn by Karel Čapek (from K. Čapek, *Letters from England* (trans. P. Selver). London: Geoffrey Bles (1925)).

capability of responding to security threats [9], security here being interpreted in its most general sense, as a threat to our survival. The disappointing uniformity of products emanating from the far-flung reaches of the British Empire was already apparent to foreign (European) visitors to the British Empire (Wembley) Exhibition of 1924 (Figure 17.1). Evidently Productive Nanosystems are *in principle* antiglobalizing, since locally adapted products can be made where. Only if subpopulations are too indolent to master the technology and produce their own designs will they lapse into a position of weakness and dependence.

17.6 An integrated approach

The message of this chapter is that there is little point in developing revolutionary nanotechnology without parallel developments in the organization of society. But the irrepressible curiosity and creativity of the scientist will inevitably drive the technology forwards—despite all the obstacles placed in the way by unsympathetic bureaucracy!—science is fundamentally a progressive activity, ever aiming at a distant goal without striving to be instantaneously comprehensive. But no similar tendency exists regarding society. The ebb and flow of social tendencies is ever-evolving and open-ended. At one level, the official vision of technology (as exemplified by policy declarations of government research councils, for example) is aimed at ever stricter scrutiny and surveillance, in which the bulk of the population is seen as a restless, unreliable mass in which criminal disturbances may break out at any instant. At another level, the spectral impalpability of so-called high finance is allowed to become ever more impenetrable and autonomous, which might be acceptable were it not for the real effects (in terms of ruined livelihoods and drastically adjusted currency exchange rates) that are now manifest.

One solid correlation that should give us undying optimism is the link between the growth of knowledge and the improvement of ethical behavior. The best hope for the future—given the impracticability of anything other than piecemeal social engineering—is to constantly promote the growth of knowledge and, given that our knowledge about the universe is still very, very incomplete, keep in mind Donald Mackay's dictum, "When data is short, keep mind open and mouth shut."

References and notes

[1] Holt GC, Ramsden JJ. Introduction to global warming. In: Ramsden JJ, Kervalishvili PJ, editors. Complexity and security. Amsterdam: IOS Press; 2008. p. 147–84.

[2] Galam S. The sociophysics of terrorism: a passive supporter percolation effect. In: Ramsden JJ, Kervalishvili PJ, editors. Complexity and security. Amsterdam: IOS Press; 2008. p. 13–37.

[3] Maxwell N. Do we need a scientific revolution? J Biol Phys Chem 2008;8:96–105.

[4] Scientometricians might argue that their work constitutes a kind of objective criticism—in principle perhaps, but in practice it degenerates into populism. For example, one of their best-known inventions is counting the number of times a published paper is cited, whence the infamous "impact factor" (the number of citations received by a journal divided by the number of papers published in the journal). Even the research director of the Institute for Scientific Information (ISI), which pioneered the extensive compilations of impact factors, recognized that they are only valid if authors scrupulously cite all papers that they should, and only those. This seldom seems to be the case. Now that the ISI has been taken over by a commercial organization (Thomson) with its own agenda, there is even less reason to put any value on an impact factor.

[5] Tran CL et al. A scoping study to identify hazard data needs for addressing the risks presented by nanoparticles and nanotubes. London: Institute of Occupational Medicine; 2005.

[6] E.g., Revell PA. The biological effects of nanoparticles. Nanotechnol Perceptions 2006;2:283–E98.

[7] Altmann J. Military nanotechnology. London: Routledge; 2006.

[8] Schweitzer A. Le problème de la paix. Nobel Lecture; 4 November 1954.

[9] Ramsden JJ, Kervalishvili PJ, editors. Complexity and security. Amsterdam: IOS Press; 2008.

CHAPTER

Ethics and Nanotechnology 18

CHAPTER OUTLINE HEAD

18.1 Risk, Hazard, and Uncertainty 204
18.2 A Rational Basis for Safety Measures 205
18.3 Should We Proceed? 206
18.4 What about Nanoethics? 207

Buckle has pointed out that the foundations of ethics have essentially not advanced for at least the last two millennia. At the same time, there have been enormous advances in what we now call human rights. Since the philosophical foundations of ethics have not changed, we must look elsewhere for the cause. What has grown spectacularly is knowledge. Therefore, it can be concluded that the reason why we treat each other on the whole much better than formerly is because we know more about the universe [1]: one should recall that one of the functions of science is to enable man to better understand his place within the universe. This advance might actually be considered the most significant contribution of science to humanity, outweighing the many contributions ministering to our comfort and convenience. Thus, in a general way, the advance of knowledge, regardless of what that knowledge is, should be beneficial to humanity.

As already pointed out in Chapter 2, knowledge may be turned into technology. Now, when we look around at all that technology has brought us, we are confronted with a familiar paradox. Explosives allow the quarryman to more expeditiously extract stone with which we can build dwellings for ourselves, but fashioned as bombs have also wrought terrible destruction on dwellings—such as in Hamburg or Dresden in the Second World War (and, more recently, in Gaza in Palestine). Nuclear fission can provide electricity without emitting carbon dioxide, but also forms the core technology of the weapons dropped with such terrible effect upon Hiroshima and Nagasaki in the Second World War; further examples seem scarcely necessary. Hence, when it is (sometimes) stated that "technology is ethically neutral," the meaning must be that there is no *net* benefit or disadvantage from its application—with some kind of ergodic principle tacitly assumed to be valid; that is, neutral provided we observe a wide enough range of human activity, or over a sufficiently long interval.

Applied Nanotechnology, Second Edition. http://dx.doi.org/10.1016/B978-1-4557-3189-3.00018-X

But, if so, then why is any technology introduced? On the contrary, technologists believe that they are embellishing life on Earth. There would be no sense in introducing any technology whose disbenefits outweigh the benefits. Hence technology is not neutral, but positive [2].

A final question for this chapter is, are ethics associated with any particular technology? Is there an ethics of the steam engine, of motoring, of cement manufacture, of space travel—and of nanotechnology? We return to this question in Section 18.4.

18.1 Risk, hazard, and uncertainty

Technological progress typically means doing new things. It may be considered unethical to proceed with any scheme that exposes the population to risk. But how much risk is acceptable? Human progress would be impossible if every step taken had zero risk. In fact, the risk of doing something for the first time is formally unquantifiable, because the effects are unknown. In practice, the variety of past experience is used to extrapolate. Steam locomotives traveling on rails allowed faster travel than previously, but initially not that much faster than a man on a horse. Tunnels caused some problems, but natural caves were known and had been explored. Flying was a greater innovation, but birds and bats and insects were familiar. Furthermore, only small numbers of people were initially involved.

New technology raises two aspects of risk: do we proceed with a new technology and do we need to regulate an already existing state of affairs? The former is typically the decision of a person or a firm. The latter is typically a collective decision of society, through its institutions.

Both cases imply firstly the need to quantify risk, and then the need to decide on a limit. Let us tackle the quantification.

Risk is usually defined as the hazard associated with an event multiplied by the probability of the event occurring. This decomposition is, indeed, the most common current basis of risk analysis and risk management. Two major difficulties immediately present themselves, however: how can hazard be quantified and how is the probability to be determined?

The answer to the first question is typically solved by using cost. Although in a particular case it might be difficult, practically speaking, because events are typically complex, nevertheless intelligible estimates can usually be made, even of the cost of events as complex as flooding. The insurance industry has even solved the problem of costing a human life. The numbers of people affected can be estimated. Hence, even if there are imponderables, a basis for estimation exists.

To answer the second question, one needs an appropriate probability model. If the event occurs reasonably frequently, the frequentist interpretation should be satisfactory. However, successive events may be correlated. Subevents aggregating to give the observed event may be additive or multiplicative, and so forth. As with the first question, heuristic approximation may be required.

The units of risk, as quantified in this manner, are cost per unit time.

Risk analysis comprises the identification of hazards (or "threats")—operational, procedural, etc.—and evidently the better one understands the operation, procedure, etc. the better one can identify the hazards. Risk management comprises attempts to diminish the hazard (its cost), or the probability of its occurrence, or both. The two are linked by comparing the costs of remedial action with the change of risk.

The overall object is to decide whether to undertake some action to diminish the hazard, or the probability of its occurrence, or both. If the cost of the action is less than the value gained by diminishing the risk, then it is reasonable to undertake the action. A slight difficulty is that sometimes a single action is carried out and there are no recurrent costs. In this case the cost of the action should be divided by the anticipated duration of its effect. A more severe difficulty is that often hazard and probability are linked. For example, installing airbags in motor-cars diminishes the hazard of an accident, but the driver, knowing this, might tend to drive more recklessly, hence increasing the probability of an accident. This factor, often neglected, frequently makes the actual effects of remedial actions very significantly less than foreseen.

The application of this kind of approach to the introduction of new technology has already been tackled in Chapter 4. In the next section, we discuss its application to regulation.

18.2 A rational basis for safety measures

The rationale behind any measure designed to increase safety is the prolongation of life expectancy, but it must do so sufficiently to prevent life quality falling as a result of the loss of income due to the need to pay for the measure.

This provides the basis for a quantitative assessment of the value of safety measures, expressed as the judgment (J)-value [3], defined as the quotient of the actual cost S of the safety measure and the maximum amount that can be spent before the *life quality index* falls.

The life quality index Q (assuming that people value leisure more highly than work) is defined as

$$Q = G^q X_d, \tag{18.1}$$

where G is average earnings (GDP per capita) from work, q is optimized work-life balance, defined as

$$q = w/(1 - w), \tag{18.2}$$

where w is the optimized average fraction of time spent working ($q = 1/7$ seems to be typical for industrialized countries), and X_d is discounted life expectancy. Note that G^q has the form of a utility function: as pointed out by D. Bernoulli, initial earnings (spent on essentials) are valued more highly than later increments (spent on luxuries). Furthermore, money available now is valued more highly than that available tomorrow; a typical discount rate is 2.5% per annum.

An individual may choose to divert a portion of his income ΔG into a safety measure that will prolong his life by an amount ΔX. Assuming ΔG and ΔX are small, expanding Eq. (18.1) and neglecting higher powers and cross-product terms yields

$$\Delta Q/Q = q\Delta G/G + \Delta X_d/X_d. \tag{18.3}$$

Since it makes no sense to spend more on safety than the equivalent benefit in terms of life prolongation, the right-hand side of Eq. (18.3) should be equal to or greater than zero. The limiting case, equality, may be solved for ΔG and multiplied by the size of the population N benefiting from the measure to yield the maximum annual sensible safety spend

$$S_{max} = -N\Delta G = (1/q)NG\Delta X_d/X_d, \tag{18.4}$$

where the minus sign explicitly expresses the reduction in income. We can then write

$$J = S/S_{max}. \tag{18.5}$$

Whether to proceed with a safety measure can therefore be decided on the basis of the J-value: if it is greater than 1, the expenditure S cannot be justified [4].

18.3 Should we proceed?

The practical ethical question confronting the entrepreneur or the board of directors of a limited liability company is whether to proceed with some development of their activities. Let us travel back to the early Victorian era. One observes that "the engineers of the Industrial Revolution spent their whole energy on devising and superintending the removal of physical obstacles to society's welfare and development" [5]. This, surely, was ethics to the highest degree. But a qualitative change subsequently occurred: "The elevation of society was lost sight of in a feverish desire to acquire money. Beneficial undertakings had been proved profitable; and it was now assumed that a business, so long as it was profitable, did not require to be proved beneficial" [5]. And there we have remained to this day, it seems. Profit has become inextricably intertwined with benefit, but the former is no guide to the latter. The utilitarian principle (the greatest benefit to the greatest number) is only useful when two courses of action are being compared. The most important principle is *elevation of society*, which should be the primary criterion for deciding whether to proceed with any innovation, without even bothering about an attempt to determine the degree of elevation—only the sign is important [6]. For Brunel, it was inconceivable that a railway engineer could have had anything other than the elevation of society in mind, hence the public was able to have total confidence in him and he could be clear-minded in his opposition to regulators. Nowadays, we have to admit that this confidence is lacking (but not, one might hope, irremediably) and, therefore, society has built up an elaborate system of regulation, which seems, however, to have hampered the innovator while providing fresh opportunities for profit to individuals without benefit to society.

18.4 What about nanoethics?

All that has been written so far in this chapter is generic, applicable to any human activity. How does nanotechnology fit in? Is there any difference between nanoethics and the ethics of the steam engine?

Possible reasons for according nanotechnology special attention are its pervasiveness, its invisibility (hence it can arrive without our knowledge), and the fact that it may be the technology that will usher in Kurzweil's Singularity.

Pervasiveness scarcely needs special consideration. All successful technologies become pervasive—printing, electricity, the personal computer, and now the internet.

The invisibility of modern technological achievement stands in sharp contrast to Victorian engineering, whose workings were generally open to see for all who cared to take an interest in such matters. Even in the first half of the twentieth century, technical knowledge was widely disseminated, and a householder wishing to provide his family with a radio, or an electric bell, would be well able to make such things himself, as well as repair the engine of a motor-car. Nowadays, perhaps because of the impracticability of intervening with a soldering iron inside a malfunctioning laptop computer or cellphone, a far smaller fraction of users of such artifacts understand how even a single logic gate is constructed, or even the concept of representing information digitally, than, formerly, the fraction of telephone users (for example) who understood how the technology worked. And even if we do understand the principle, we are mostly powerless to intervene. Genetic engineering is, in a sense, as familiar as the crossing of varieties known to any gardener, but few people in the developed world nowadays grow their own vegetables [7], yet there is mounting frustration at the unsatisfactory quality, in the culinary and gastronomic sense, of commercially available produce.

The answer to invisibility must, however, surely be a wider dissemination of technical knowledge. This should become as pervasive as basic literacy and numeracy. Hence it needs to be addressed in schools. It should be felt to be unacceptable that even basic concepts such as atoms or molecules are not held as widely as knowledge of words and numbers. Today they are not; to make "technical illiteracy" a thing of the past will require a revolution in educational practice as far reaching as the Nano Revolution itself. To be sure, just as among the population there are different levels of literacy and numeracy, so we can expect that there will be different levels of technical literacy, but "blinding by science" should become far more difficult than it is at present.

Even without, or prior to the occurrence of, the Singularity, Productive Nanosystems (PNs) imply an unprecedented level of technological assistance to human endeavor. All technologies tend to do this, and in consequence jobs and livelihoods are threatened—we have already mentioned Thimonnier's difficulties with tailors. The traditional response of engineers is that new jobs are created in other sectors. This may not be the case with PNs, in which case we can anticipate a dramatic shift of the "work–life balance" q in favor of leisure (see Section 18.2). Provided material challenges to human survival (Chapter 17) are overcome, finding worthwhile uses of this extra leisure remains as the principal personal and social challenge. Here it should be noted that as work–life balance shifts in favor of life (i.e., more leisure), implying decreasing q [8], the marginal utility of money possessed saturates much more rapidly (Eq. 18.1). This may have profound social consequences (involving greed or its absence [9]), which have not hitherto been analyzed and which are difficult to predict.

The answer should, however, be available within the world of Productive Nanosystems (Section 16.1), with which everybody will be able to create his or her own personal environment. It represents the ultimate control over matter, and does not depend on an elite corps active behind the scenes to maintain everything in working order. In this view, the age-old difference between the "sage" and the "common people" that is taken for granted in ancient writings such as the *Daodejing* should disappear. Is such a world, which each one of us can shape according to our interests and abilities, possible? That is at least as difficult a question to answer as whether Productive Nanosystems will be realized.

Even if they are, and even if we all become "shapers," will not our different ideas about shaping conflict with each other? Will there not be incompatible overlaps? That is presumably why we shall still need ethics. But human solidarity should be enhanced, not diminished, by more knowledge of the world around us, and it is that to which we should aspire. Anything less represents regression and loss. Let us proceed with nanotechnology in this spirit, not indeed knowing whither it will lead, but holding fast to the idea of elevation.

References and notes

[1] As Buckle would have said, moral truths are stationary, and dependent on the state of intellectual knowledge for their interpretation. Buckle HT. History of civilization in England, vol. 1. New York: D. Appleton & Co.; 1864. p. 130 [chapter IV].

[2] In some cases, it transpires that what is a benefit to one subpopulation is a disbenefit to another (the latter usually having no say about the development). According to the principle of human solidarity, this is unethical.

[3] Thomas PJ et al. The extent of regulatory consensus on health and safety expenditure. Part 1: Development of the J-value technique and evaluation of regulators' recommendations. Trans IchemE, Part B 2006;84:329–36.

[4] Thomas PJ et al. The extent of regulatory consensus on health and safety expenditure. Part 2: Applying the J-value technique to case studies across industries. Trans IchemE, Part B 2006;84:337–43.

[5] Weir A. The historical basis of modern Europe. London: Swan, Sonnenschein, Lowrey; 1886. p. 393–4.

[6] It is a telling difference that it is quite typical nowadays for visionary and unexceptionably beneficial projects to be associated with the names of their business promoters, who may have nothing to do with the intellectual achievement of the innovation *per se*, the names of the engineers remaining unknown to the public, whereas in the Victorian age the opposite was the case.

[7] Even if they did, they might find it difficult to procure seeds corresponding to their desires.

[8] Note that the simple equation (18.2) cannot be used directly here to compute q from a new w, because it deals with optimized quantities.

[9] Ramsden JJ. Psychological, social, economic and political aspects of security. In: Ramsden JJ, Kervalishvili PJ, editors. Complexity and security. Amsterdam: IOS Press; 2008. p. 351–68.

Further reading

[1] Allhoff F, Lin P, Moor J, Weckert J, editors. Nanoethics. Wiley; 2008.
[2] Groves C. The public perception of nanotechnology: is it all about risk? Nanotechnol Perceptions 2010;6:85–94.
[3] Ramsden JJ, Aida S, Kakabadse A, editors. Spiritual motivation: new thinking for business and management. Basingstoke: Palgrave Macmillan; 2007.

Epilog

One of the key themes of our time is sustainability, which in its widest sense means the ability of human civilization to continue to exist and to develop. Nanotechnology provides some of the means to achieve sustainability, so is South Korea, whose dramatically impressive progress in nanotechnology has been noted, an example for us all to follow? On the other hand, the country is said to have tremendous social problems as evinced by the rapidly falling birthrate. But, painful as that is in the short term (as other countries are finding too) in the longer term the whole world has to drastically reduce its population. Without such a reduction it is very difficult to foresee how nanotechnology, however minimal the resources it requires and however maximal the functionality it can confer on materials and devices, can rescue a rapidly degrading environment.

In the immediate term, nanotechnology is clearly a natural continuation of the trend toward ever-improving control of structure at all length scales. Because it can so often be initially applied in a substitutional fashion, it also satisfies the practical industrialist's need to minimize the introduction of brand-new materials.

Nanotechnology needs, of course, to be applied responsibly. This includes open and transparent trading to agreed standards. Ultimately, it is hoped that all nanotechnologists will make their personal integrity the first consideration, from which responsibility and the rest follows.

Index

A

AAGR. *See* Average annual growth rate (AAGR)
Accumulated value, 20
Active nanodevices, 190
Adenosine triphosphate (ATP), 84
Advanced Technology Centre (ATC), 172
Aerospace, and automotive industries, 66
Agile manufacturing, 44
Aim-oriented empiricism (AOE), 198
Alkali Inspectorate, 153
"Alternative model", 21–22
Amplification, 29
"Anti-graffiti" paint, 66
Architecture, and construction, 66
Atom-by-atom assembly, 68
Atomically precise technologies, corollary of, 68
Atomic force microscope (AFM), 4
Automatic diagnosis, diseases, 107
Automotive industries, aerospace and, 66
Average annual growth rate (AAGR), 52

B

Bayh-Dole Act, 27
Benign dictatorship, 118
Bionanotechnology, 9
Business environment
  clusters, 120
  company-university collaboration, 119
  development timescales, predicting, 125
  intellectual needs, 117
  nanometrology, 127
  nanotechnology, assessing demand for, 120
    anticipating benefit, 122
    innovation value, 121
    modeling, 121
  nanotechnology, radical nature of, 116
  nanotechnology, universality of, 113
  patents, 130
  standardization of, 129
  technical and commercial readiness levels, 123

C

Cambridge Display Technology (CDT), 165
Capture value, 115
Carbon-based materials, 63
Carbon nanotube arrays, 99
Catalysis, 67
Caveat emptor, principle of, 59
CDT. *See* Cambridge Display Technology (CDT)
Centre National pour la Recherche Scientifique (CNRS), 172
Civilization Index (CI), 173
Classification, Labelling and Packaging (CLP), 154
Climate change, 195
CMOS. *See* Complementary metal oxide-semiconductor (CMOS)
CNRS. *See* Centre National pour la Recherche Scientifique (CNRS)
Commercialization, costs of, 116
Complementary metal oxide-semiconductor (CMOS), 98
Complex machines, 190
Composite films, 74
Computer modeling, 126
Concept of regulation, 153
Conceptual nanotechnology, 9
Conditional knowledge, 13
"Core-shell" particle, 67
Crowdsourcing, 179
Customized pharmaceuticals, 107

D

Data storage technologies, 99
Decree-driven model, 14
Demography, 196
Depletion, of resources, 197
Design, improvements in, 39
Design algorithm, 178
Diagnostic imaging applications, 105
Direct nanotechnology, 9
Dispassionate observer, 58
Display technologies, 100
Double counting, 51
Downstream audit sequencing (DAS), 155

E

ECM. *See* Extracellular matrix (ECM)
Economic forecasting, traditional methods of, 185
Electrical cabling, 92
Electrical storage devices, 88
Elevation of society, 206
Enact pharma, 163

Energy
 efficiency, 91
  electrical cabling, 92
  lighting, 92
 harvesting, 84
 localized manufacture, 94
 production, 85
  fuel cells, 86
  oil and gas industry, nanotechnology in, 87
  solar energy, 85
 storage
  electrical storage devices, 88
  hydrogen storage, 90
Engineering, *vs.* science, 186
ENIAC computer, 18
Environment, 67
Environmental degradation, 196–197
Environmental Protection Agency (EPA), 154
Ethics
 nanoethics, 207
 risk/hazard and uncertainty, 204
 safety measures, rational basis for, 205
European Union "Framework" research, 139–140
Extracellular matrix (ECM), 106

F

Fabricating nanofibers, general-purpose technology for, 78–79
Fabrication, 42
Fabrication, performance, 43
Fast ion bombardment (FIB), 128
Field-asymmetric ion mobility spectrometry (FAIMS), 166
Field Studies Council (FSC), 189
Fiscal environment
 endogenous funding, 141
 government funding, 141
 nanotechnology funding, geographical differences, 148
 sources of funds, 137
Food
 consumer choice, 72
 farming, 70
 nano-additives, 71
 nanotechnology and crisis, 73
 packaging, 69
 sensors, 70
 social context, 72
Food chains, 68
Forward selling products, 138, 146
Fuel cells, 86

G

Genetic algorithms, 178
Genetically modified (GM) plants, 33–34
Genetic engineering, 207
Genetic programming, 178
Giant magnetoresistance (GMR), 99
Globalization, 32, 199
Global market value, 52
GMR. *See* Giant magnetoresistance (GMR)
Gold nanoparticles, 104
Government funds, 139–142
Gratis, 179–180

H

Health
 current activity
  information technology, 106
  nanomaterials, 106
  nano-objects, 104
 implanted devices, 107
 longer-term trends, 107
 paramedicine, 109
Heat management, 98
Homeland security, 77
Hydrogen economy, 87
Hydrogen storages, 90
Hyperion, 164
Hyperion Catalysis International, 165

I

Indirect nanotechnology, 9
Indium-doped tin oxide (ITO), 100
Information technologies
 data storage technologies, 99
 display technologies, 100
 heat management, 98
 Molecule/particle sensing technologies, 100
 silicon microelectronics, 98
Innovation
 classification of, 30
 creative destruction, 29
 development of, 32
 management, 33
 maturity, effect of, 34
 time course of, 27
Integrated approach, 200
Integrated Nano-Science and Commodity Exchange (INSCX), 173
Internal funds, of company, 138

International Monetary Fund (IMF), 141
International Standards Organization (ISO), 3
Iron-containing nanoparticles, 67

K

Key enabling technologies (KETs), 154
K-limited régime, 29

L

Life quality index, 205
Lighting, 92
Lightweight refractory nanocomposites, 88
Linear Baconian model, 117
Linear model, 15, 17–18
"Linear model", 25–26
Lithotrophic prokaryotes, 76
"living proof of principle", 103
Long-term implants, 108
Low-friction coating, 74
Lubricants, 74

M

Magnetic fluid hyperthermia, 105
Magnetic tunnel junction (MTJ), 99
Manufacturability, 125
Manufacturing systems, miniaturization of, 41
Market pull, 27
Material safety data sheet (MSDS), 154
Materials selection, 180
Mechanical movement, 84
Merging materials research, 33
MesoPhotonics Ltd., 162
Metal extraction, minerals, 75
Metal nanomaterials, 53
  range of applications for, 54
Metal oxide-semiconductor transistor, 27–28
Metrology
  instrumentation, 75
  nanotechnology and, 129
Microdiversity, recognition of, 29
Micro electromechanical systems (MEMS), 84
Microprocessor, development of, 97
Microsystems technologies (MST), 84
Minerals, and metal extraction, 75
Miniature processor-enabled personal computer, 41
Miniaturization
  dramatic progress in, 40
  enabled personal computer, 41
  of manufacturing systems, 41
Molecular detection efficiency, 108
Molecular electronics, 191
Molecule sensing technologies, 100

N

Nanobiotechnology, 51
NanoCo Technologies Ltd., 164
Nano-engineered "artificial kidneys", 76
NanoMagnetics Ltd., 161–162
Nanomaterials market, 51
Nanomaterials segment, 53
Nanomedicine, 103
Nanometrology, 127
Nano-objects, addition of, 78
Nanoparticles
  biological effects, 57
  conferring, 78
Nano revolution, 127
Nanoscale, 10
Nanoscale metal oxide nanoparticles, 53
Nanoscience, 11
Nanoscopic robots, 105–106
Nanostructured coatings, 91
Nanostructured materials, 113
Nanotechnology, 17, 51
  and bionanotechnology, 9
  definition, 10
  in developing world, 174
  as devices and systems, 8
  direct/indirect and conceptual, 9
  evolution of, 18
  and food crisis, 73
  geography of
    countries categories, 173
    locating research partners, 171
    locating supply partners, 173
  main application for, 66
  as materials, 7–8
  nanoscale, 10
  nanoscience, 11
  as process, 4
  specific challenges, 199
Nanotechnology Availability Levels (NAL), 123–124
Nanotechnology business
  consumer products, 55
  current situation of, 52
  nanoproducts, safety of, 57
  products, types of, 54
  statistics, 49
  total market of, 50

Nanotechnology companies
CDT, 165
Enact pharma, 163
generic business models, 167
Hyperion, 164
MesoPhotonics Ltd., 162
NanoCo Technologies Ltd., 164
NanoMagnetics Ltd., 161
Owlstone, 166
Oxonica, 163
Q-Flo, 166
Nanotechnology-enabled products
global market for, 52
Nanotechnology funding
and geographical differences, 148
Nanotechnology products
design of
crowdsourcing, 179
enhancing traditional design routes, 178
materials selection, 180
vastification, challenge of, 177
incrementally improved, 54
radically new products, 55
of substitution, 54
Nanotools, 52
Nara Institute of Science and Technology (NAIST), 161
Natural philosophy, 15
Natural photosynthesis, 85
"Near-frictionless carbon" (NFC), 75
"New model", 17–18, 22, 25
Noncarbon materials
coatings, 63
composites, 62
Nonpolymer organic materials, 53

O

Occupational Health and Safety Act (OSHA), 154
OEM. *See* Original equipment manufacturer (OEM)
OLEDs. *See* Organic light-emitting diodes (OLEDs)
Organic light-emitting diode (OLED), 165
Organic light-emitting diodes (OLEDs), 100
Original equipment manufacturer (OEM), 114
Outsourcing, 179
Owlstone, 166
Oxonica, 163

P

Paper, 76
Parallelization, 42
Particle sensing technologies, 100
Pervasiveness, 207
Polymer manufacturer, 147
Precision, 42
Precoating surfaces, alternative to, 74
Prey (Crichton, Michael), 189
Private investors, 138
Productive nanosystems, 179–180, 186–187, 190, 199
social impacts, 188
technology
carbon-based feedstock, 188
individualized production, 187
localized production, 187
production on demand, 187
ultralow-cost production, 187
timescales, 187
Productive nanosystems (PN), 122, 207
Prokaryotic microbes, 76
Punctuated equilibrium, 30

Q

Q-Flo, 166
Quantitative indicators, 50
Quantum computing, 192

R

Radioactivity, and atomic fission, 17–18
Registration, Evaluation and Authorization of Chemicals (REACH), 154
Reliability, 43
Rensselaer Polytechnic Institute (NY), 172
Reproducibility, 43
Ribonucleic acid (RNA), 27
Risk analysis, 205
Royal National Lifeboat Institution, 143

S

Safety, health, and environment (SHE), 147
Scaling, 41
Scanning electrochemical microscopy (SECM), 128
Scanning ion current microscopy (SICM), 128
Scanning probe microscope (SPM), 5

Scanning tunneling microscope (STM), 4
Science
  engineering *vs.*, 186
  social value of, 21
  technology emerging from, 25
Scientific research, public funding of, 143
Second World War, 203
Security, 77
Seiko-Epson Corp. (SEC), 165
Self-assembly, 190
Selfridge, 119
Semiconductor industry, 146
Sensorization, 107
Shapers, 208
Silicon microelectronics, 98
Small Is Beautiful (Schumacher), 189
Society's grand challenges
  material crises, 195
  science crisis, 198
  social crises, 198
Soil remediation, 67
Solar energy, 85
Standard empiricism (SE), 198
Standardized items, 145
Steam locomotives traveling, 204
Storing energy, 87
"Subnano" technologies, 74
Substance-sensing devices, 108
Supercapacitors, 87, 89

T

Technical specification (TS), 3
Technological infrastructure, 172
Technology, evolution of, 18
Technology Readiness Level (TRL), 123
Terrorism, 197
Textiles, 77
The Advancement of Learning (Bacon, Francis), 14, 21–22
Thermal interface materials, 98
Topografiner, 4
Toxic Substances Control Act (TSCA), 154
Traceability, 69
Traditional technology, 188
Traditional tribopairs, 108
Trend ranking, 185
TRL. *See* Technology Readiness Level (TRL)
Tunnel magnetoresistance (TMR), 99

U

Ultraprecision engineering, 65
Ultrathin film coatings, 66
Unconditional knowledge, 13
Uniformity, 43
Universal technologies, 114
Unsourcing, 180
US Federal Food, Medicine, and Cosmetic Act, 154

V

"Value in use", 20

W

Wealth, nature and value of, 20
Wear, 42
Woodrow Wilson Center, 55

Printed and bound by CPI Group (UK) Ltd, Croydon, CR0 4YY
08/05/2025
01864838-0005